AIRCRAFT SHEET METAL

DRAKE'S AIRCRAFT MECHANIC SERIES

AIRCRAFT WOODWORK

AIRCRAFT WELDING

AIRCRAFT SHEET METAL

AIRCRAFT ENGINES

AIRCRAFT ELECTRICAL SYSTEMS,
HYDRAULIC SYSTEMS,
AND INSTRUMENTS

AIRCRAFT MAINTENANCE AND SERVICE

AIRCRAFT ENGINE MAINTENANCE AND SERVICE

AIRCRAFT SHEET METAL

by COLONEL ROLLEN H. DRAKE, B. S., M. A.

RESEARCH ENGINEER, PILOT AND GROUND SCHOOL INSTRUCTOR
FORMERLY: INSTRUCTOR, VOCATIONAL AND AVIATION SUBJECTS, LOS ANGELES CITY SCHOOLS; CHIEF, AIR AGENCY UNIT AND CIVILIAN PILOT TRAINING SPECIALIST, CIVIL AERONAUTICS ADMINISTRATION; SUPPLY SPECIALIST, OFFICE OF THE QUARTERMASTER GENERAL

Originally published in 1947

ISBN: 978-1-940001-36-4

The Aviation Collection
by
Sportsman's Vintage Press
2015

This book is dedicated to

MY BELOVED WIFE

without whose inspiration and help

this book could not have been written

PREFACE

FROM earliest time man has used metal in the form of sheets. Crude lumps of native metal could not be readily used, so he hammered it into sheets to be used in utensils, armor, and weapons. There is hardly an industry of today which does not make use of sheet metal. This is particularly true in the field of aircraft construction. This book is written primarily for classroom use but it will also serve as a practical manual and guide for the man in the field. It covers all phases of aircraft sheet metal work and contains information on the characteristics and uses of metals employed in aircraft fabrication. Particular attention is given to the fundamental theory of forming and fabricating metals into sheet form. Included also is considerable information on the various metals and their alloys and their treatment before and during fabrication.

All the theory necessary to meet the sheet metal requirements of an aircraft mechanic's license is included. It will meet the requirements for college vocational courses, trade schools, junior colleges, high schools, and aviation ground schools. It is written in nontechnical simple language. The author avoids the use of formulas, graphs, confusing tables, obscure footnotes, and any other material which will not be clearly understood by everyone.

The various methods of forming and fastening sheet metal have been explained in detail.

Emphasis has been placed on forming and joining of sheet metal and also on the detection of faulty workmanship and other common faults occurring in this kind of work. The different metals and their alloys are treated separately with detailed information on forming and fabrication of each group. Emphasis is placed on types of approved repairs required by the Civil Aeronautics Administration. Proper methods of making these approved repairs are thoroughly explained and illustrated. Since the average student usually has great difficulty in getting a clear picture of the mechanical processes or operations from printed material without visual help, the text is illustrated fully and in great detail.

The author wishes to express his grateful appreciation to the following who have so kindly furnished material which has been of great assistance in the preparation of this book: the United States Office of Education; Civil Aeronautics Administration; the Department of Education of the various states, particularly New York, Pennsylvania, Utah, and Virginia; Douglas Aircraft Company; Piper Aircraft Corporation; Curtiss-Wright Corporation; Timm Aircraft Corporation; Glenn L. Martin Company; Reynolds Metals Company; Bell Aircraft Corporation; Aluminum Company of America; Dow Chemical Company; Cleveland Twist Drill Company; Cleveland Pneumatic Tool Company; High Speed Hammer Company; and to Howard E. Hunter, Foreman, Airplane Overhaul Shop, Intercontinental Division, Transcontinental and Western Air Incorporated, for the reading of, criticism of and suggested changes in the manuscript.

The photographs by Al Palmer, released by the Office of War Information, are by courtesy of Allegheny Ludlum Steel Corporation.

The author wishes to thank his many friends who have so generously contributed their advice and assistance. He also wishes particularly to express his gratitude to Earle R. Hough for his excellent work in the preparation of many of the drawings in this book, and to Mildred Pickrel and Alma Franklin for their untiring patience and cooperation in the preparation of the manuscript.

The frontispiece shows the forming of an aluminum alloy aircraft part with a drop hammer, and is reproduced by courtesy of the Glenn L. Martin Company.

R. H. D.

TABLE OF CONTENTS

AIRCRAFT SHEET METAL

1 INTRODUCTION

Man from prehistoric times has hammered metal into sheets. Early man, of course, used only those metals which were found native. Most metals are not found in nature in their pure state, but are combined with other substances in the form of chemical compounds. These compounds must be refined by smelting or other processes to obtain the pure metal. Copper and gold are found native. Copper, being ductile and easily worked, and found in comparatively large quantities in the pure state, was first used by early man for tools and utensils. The use of copper was so extensive following the Stone Age that this period of history has been called the "Copper Age."

Tin in the form of its oxide is quite easy to refine and, when alloyed with copper, forms bronze which was the important metal used throughout the Bronze Age. Bronze was hammered into sheets by hand and used for utensils and armor.

The discoveries of methods of obtaining iron from its ores by smelting led to the use of iron instead of bronze. Iron was more plentiful and, when alloyed with carbon, was stronger and harder than copper or bronze.

The metal worker has hammered metals into sheets for hundreds of years, but it was not until the development of heavy machinery for the working of metal that sheet metal could be produced economically and in large quantities. This led to the use of sheet metal throughout industry. The major portion of all metals produced is furnished to the consumer in the form of sheet and plate.

A substantial percentage of the aluminum produced today is used in aircraft manufacture. The alloys of steel are also used throughout the aircraft industry. One of the large new transport airplanes is covered with stainless steel of a thickness no greater than a sheet of paper. Magnesium and its alloys are coming into use because of their high strength-weight ratio.

Most of the sheet metal used in industry is formed by the aid of machinery. Handwork by the skilled sheet-metal worker is, however, necessary to its most efficient use. Not only does the sheet-metal worker direct the work of the machine but, particularly in the aircraft industry, it is necessary for him to make many of the fine adjustments by hand. Almost every part of an airplane which is formed by machinery must have some hand forming done upon it before it is ready for final installation in the airplane. Many parts, even in the most modern aircraft factory, are formed by hand. It is impossible for machinery to perform much of the work done on small irregular parts. The tools of the sheet-metal worker vary from giant presses weighing many tons to delicate

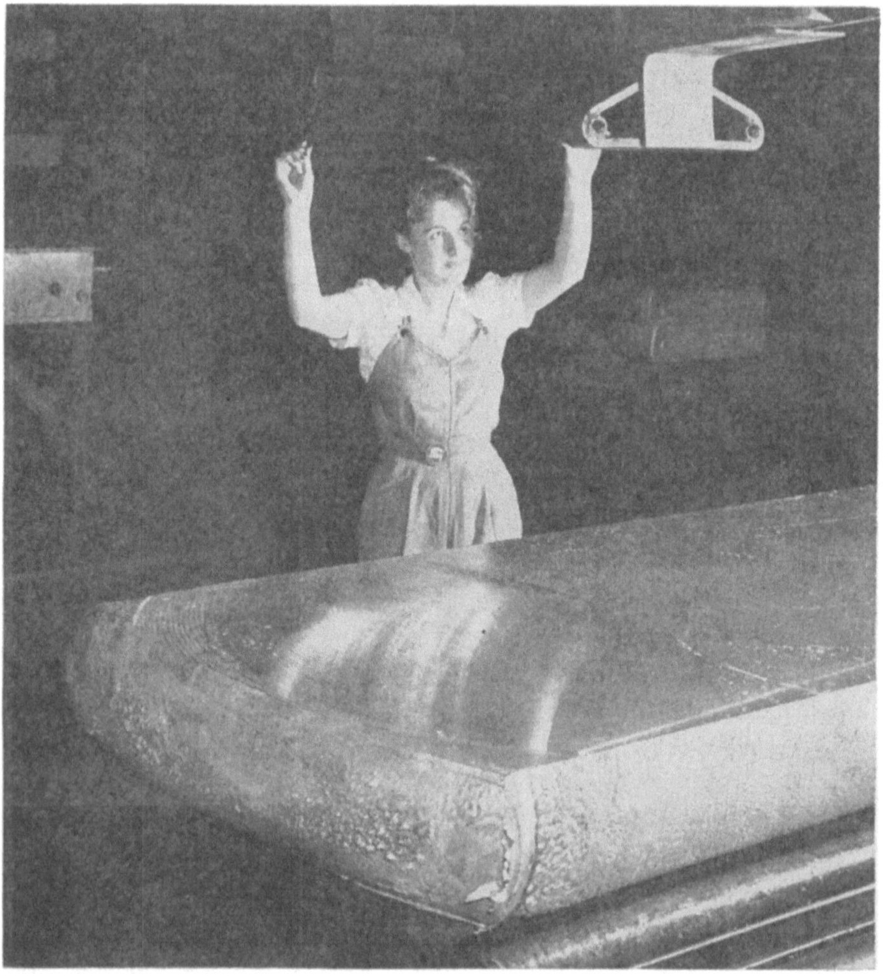

Fig. 1. An ingot of aluminum alloy before rolling. (Courtesy Reynolds Metals Company)

Fig. 2. An ingot of aluminum alloy rolled into thin sheet. (Courtesy Reynolds Metals Company)

Fig. 3. A huge hydraulic press used to form aircraft parts. (Courtesy Douglas Aircraft Company, Inc.)

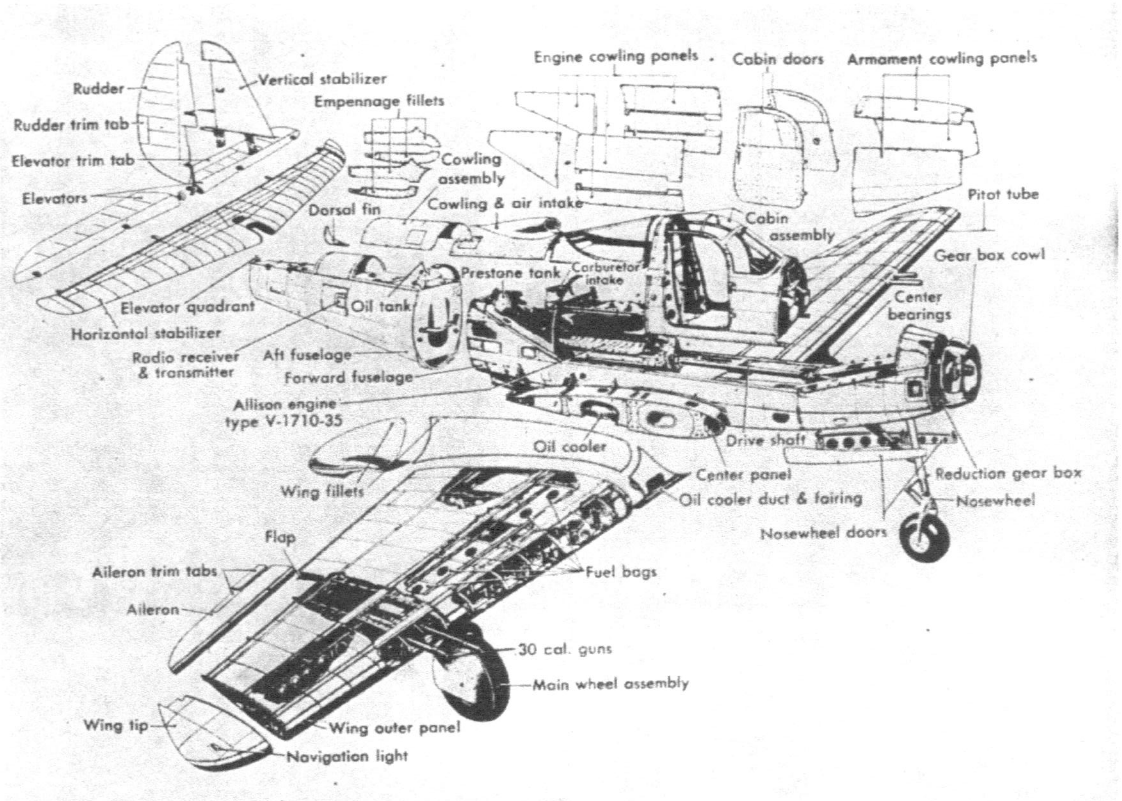

Fig. 4. An exploded view of an all-metal aircraft. (Courtesy Bell Aircraft Corporation)

hammers and tools weighing but a few ounces. Most of the handwork requires fine judgment and a high degree of skill.

It is necessary for the machine operator to have an understanding of the behavior of metals while they are being worked. He should understand the principles of stretching, shrinking, drawing, and cold-working to obtain the greatest efficiency from his machine. A thorough understanding of these processes can be had only through the working of metal by hand. The skilled sheet-metal worker controls the flow and movement of the metal and can make it do what he wishes.

Aircraft Nomenclature. Before beginning actual work in aircraft sheet metal it is necessary to become familiar with the various parts of the airplane structure. The main divisions of an airplane are: (1) the body of the airplane, called the fuselage; (2) the airfoils, or wings, which, through their reaction to the air, support the weight of the airplane in flight; (3) the tail group, called the empennage, consisting of the rudder, elevators, vertical fin, and horizontal stabilizer; (4) the landing gear, which includes the wheels, tail wheel, shock absorbers, and the necessary bracing; and (5) the engine mount, which is the main support of the engine, and the necessary cowling to enclose the engine.

The fuselage of all airplanes is similar in design and location. The main differences are the size and the use for which the airplane is designed. A small, light airplane has a fuselage arranged to carry one or two passengers, including the pilot, and a few pounds of baggage, which make up the entire payload of the airplane. The large transport planes have rows of seats arranged to carry many passengers with room in the front of the fuselage for baggage, radio equipment, etc. In this type of airplane, the pilot and copilot are far forward in the fuselage in a separate compartment from the passengers. The fuselage of cargo-carrying airplanes has large spaces in which cargo may be stored.

In light airplanes, the fuselage may be made of welded steel tubing covered with fabric, a wooden framework covered with light plywood, a combination of metal and plywood, of wood and fabric, or of sheet metal.

The wings are much the same for every type of airplane. The method of construction varies much more widely than the shape. In many military aircraft, large transport, and cargo airplanes, the wings have no external support and are made almost entirely of metal. This type of construction is called "cantilever."

In light airplanes, the main fuselage structure, engine mount, landing gear, and control surfaces are usually of welded metal construction.

Fig. 5. A modern light airplane. (Courtesy Piper Aircraft Corporation)

Fig. 6. A modern transport flying boat. (Courtesy The Glenn L. Martin Company)

The fuel tanks of most airplanes are located in the wings. The design of the wing depends largely upon the type of airplane and the purpose for which the airplane is to be used.

The tail group, or empennage, is designed almost entirely to give control to the airplane in flight and acts somewhat like the feathers on an arrow in maintaining the direction of flight. The movable portions,

Fig. 7. A modern all-metal military aircraft. (Courtesy The Glenn L. Martin Company)

namely, the rudder and elevators, make it possible for the pilot to change the direction of flight.

The landing gears of practically all aircraft are made of metal because of the severe strains placed upon them in taxiing over rough ground, landing, and taking off. Rubber tires, rubber shock cords, and air and oil cylinders are used to reduce shocks to the airplane structure.

When the engine is not mounted in front of the fuselage, the housing which encloses the engine and its various accessories is called a nacelle. The engine mounts of airplanes are strong, rigid structures, usually of metal. This is necessary because of the strains placed upon the mount by the engine which is turning the propeller at a high rate of speed.

II GENERAL PROPERTIES OF METALS

It is necessary for the sheet-metal worker to have a good knowledge of metals in order to understand thoroughly the materials with which he works. The worker in sheet metal is not expected to know all there is to know about metals and their alloys. This is a large field and a study within itself. However, a good understanding of the physical properties of the common metals and their alloys is necessary, as is a knowledge of the effect of heat treatment and working on these properties.

Metals, when pure, are classed as elements. There are 92 elements in all, roughly divided by the chemist into metals and nonmetals. The metals all have "metallic luster" on freshly polished surfaces. Most of the metals combine readily with other elements, forming chemical compounds. When a metal combines with oxygen, the compound formed is called an oxide.

The metals differ greatly in their physical properties. Physical properties are those properties which can be determined by the senses, such as color, weight, strength, and hardness.

The weight per unit volume of metals, which is called density, varies widely. For example, lithium weighs 33.072 lb. per cu. ft., and a cubic foot of iridium weighs 1397.76 lb.

The specific gravity of any substance is the number of times a volume of that substance is as heavy as an equal volume of water. For example, a cubic inch of lead weighs 11.34 times as much as a cubic inch of water. Therefore, the specific gravity of lead is 11.34. The specific gravity of aluminum is 2.7; of gold, 19.3.

Metals vary in their hardness. Lead is one of the soft metals. Sodium is lighter than water and is almost as soft as putty. Some of the metals are so hard that they will scratch glass. Some are brittle and break easily when pounded. Others are soft and may be pounded into any shape without breaking. Gold may be hammered into sheets so thin that the sheet is almost transparent.

The melting point of metals, which is the temperature at which they change from the solid to the liquid state, covers a wide range. Lithium melts at about 367° F.; lead melts at 621° F.; while the melting point of tungsten is 6098° F. Aluminum melts at 1220° F.; and steels have melting points from 2480° F. to 2786° F., depending upon the alloy.

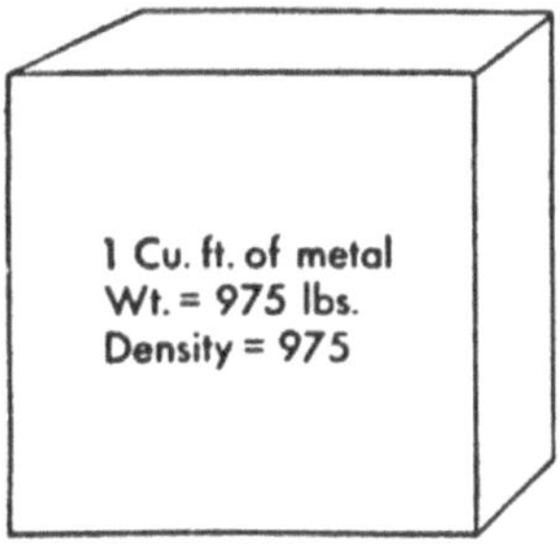

Fig. 8. Density equals the weight per unit volume. The density of a substance is commonly given in pounds per cubic foot.

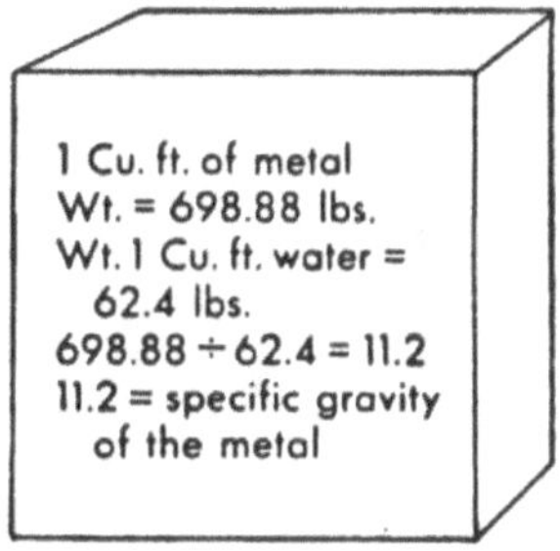

Fig. 9. The specific gravity of a substance equals the weight of a unit volume of the substance divided by the weight of an equal volume of water.

All metals conduct electricity. Silver has the highest conductivity of all the metals, and its conductivity is taken as 100. The conductivity of bismuth is only 1.8 in comparison with that of silver.

All metals except copper and gold are silvery white in color when freshly polished; copper has a reddish color, and gold is yellow.

Some metals corrode readily, while others, such as gold, silver and platinum, corrode hardly at all under ordinary conditions. Aluminum corrodes rapidly. A layer of aluminum oxide, which corresponds to iron rust, forms on freshly polished aluminum within a few minutes. Corrosion is sometimes called tarnish.

The volume of a piece of metal changes with changes in temperature. All metals expand when heated and contract when cooled. Almost all of the metals contract when passing from the liquid to the solid form. Each metal has its own rate of expansion. The amount which a metal changes in length or volume for each degree of change in temperature is called the "coefficient of expansion." When expansion is measured lengthwise, it is called "linear expansion per inch per degree." The amount the metal expands or contracts for each degree change in temperature is very small. For example, iron expands 0.00000648 in. for each degree change in temperature. This means that a piece of iron wire 1 in. long expands or contracts 0.00000648 in. for each degree change in temperature. Aluminum, lead, tin, and zinc have a comparatively large coefficient of expansion. The coefficient of expansion of aluminum is 0.00001234; of magnesium, 0.0000143. After metal has been caused to expand by heating, it always returns to the original dimension when cooled to the original temperature.

Fig. 10. Aluminum sheet is furnished to the aircraft manufacturer in large rolls. (Courtesy Douglas Aircraft Company, Inc.)

The electrical conductivity of metals changes with changes of temperature. The conductivity of metal is lower at high temperatures than at low temperatures.

Few metals are used in their pure state. Copper, gold, platinum, and tin, however, are used in the pure state. Much copper in a highly refined form is used in the manufacture of electrical equipment. Considerable quantities of commercially pure aluminum are used in industry. Several of the metals, such as tin, chromium, and nickel, are used in the pure form to coat other metals in order to increase their resistance to corrosion.

Most metals are used in the form of alloys. An alloy is formed by melting two or more metals together. Metals dissolve in other metals in much the same way that sugar dissolves in water. If sugar is dissolved in water and the water is frozen, a sweet ice is formed, which would represent an alloy of sugar and water. Steel is considered to be an alloy of carbon and iron.

A true alloy is a solution of metals within metals, or a mixture of metals. Carbon is not a metal. There are a number of elements other

Fig. 11. Extruded tubing and structural parts. (Courtesy Reynolds Metals Company)

than metals, such as carbon, phosphorus, and sulphur, which are commonly found in steels and which affect the properties of the steel. These elements, which are nonmetals, form compounds with iron or may be mixed with iron. While these substances do not form true alloys, the properties of the alloys are often affected to a considerable extent by their presence. Even very small quantities of a metal, or these nonmetallic substances, may have a great effect on the properties of the alloys formed. For example, a fraction of 1 per cent of sulphur may make steel unfit for use in aircraft construction.

Commercially, many metals are classified according to the method of preparation. When the metal is poured into a mold the metal is said to be "cast." Metal may be hammered or forged, rolled or extruded. Extruded metals are products which have been squeezed through variously shaped holes under heavy pressure.

There are a number of physical characteristics of metals which should be understood by the sheet-metal worker.

Hardness. The hardness of the metal is that property by which it resists scratching, cutting, denting, or abrasion.

Brittleness. Brittleness is that property of metal which causes it to break when bent, deformed, or hammered.

Fig. 12. Hardness.

Fig. 13. Brittleness.

Fig. 14. Malleability.

Malleability. The malleability of a metal is that property which allows it to be formed, bent, or hammered into thin sheets without breaking.

Ductility. Ductility is similar to malleability and is that property of a metal which allows it to be drawn into wires without breaking.

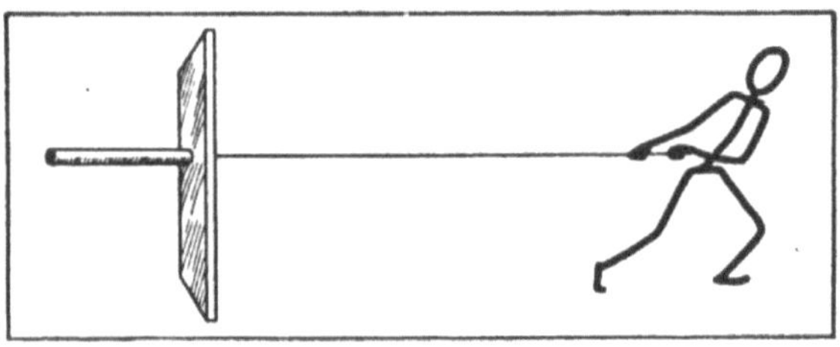

Fig. 15. Ductility.

Elasticity. Elasticity is that property of a metal which causes it to return to its original shape after being bent, compressed, or stretched.

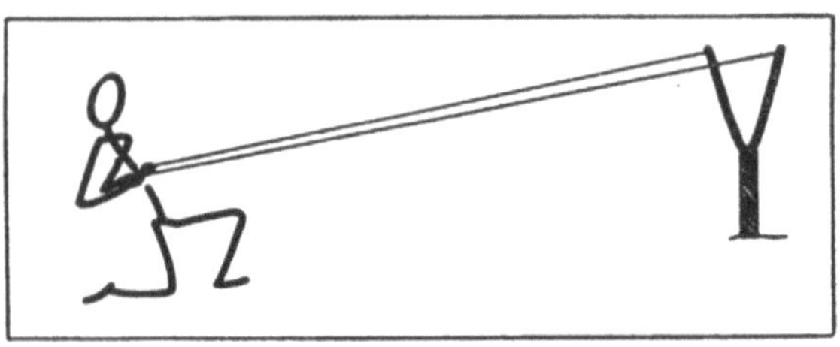

Fig. 16. Elasticity.

Toughness. Toughness is that property of a metal which allows it to be deformed without breaking.

Heat Conductivity. Heat conductivity is the rate at which heat is conducted through the metal.

Tensile Strength. Tensile strength is that property of the metal which resists pulling forces such as would tend to pull the metal apart.

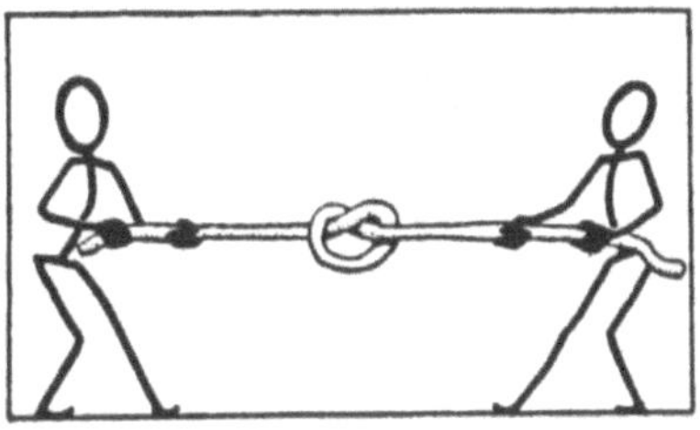

Fig. 17. Toughness.

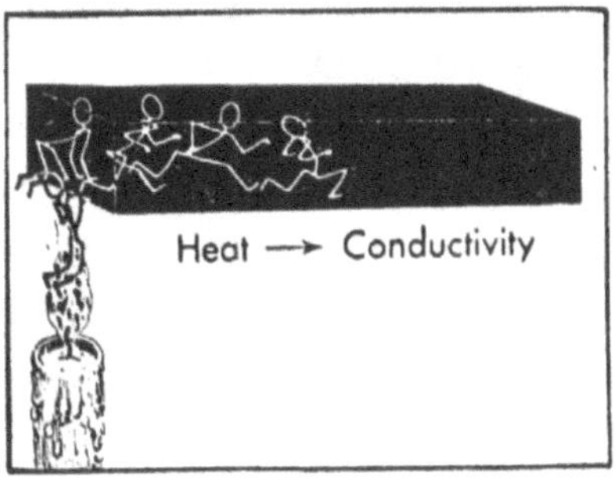

Fig. 18. Heat conductivity.

Compression Strength. Compression strength is that property of the metal which resists crushing forces or forces which tend to squeeze the metal together.

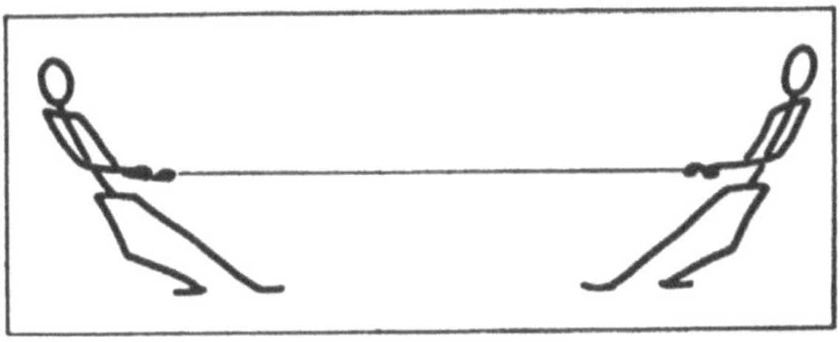

Fig. 19. Tensile strength.

Fig. 20. Compression strength.

Shearing Strength. Shearing strength is that property of the metal which resists forces which tend to cut it in the manner of a pair of shears.

Fig. 21. Shearing strength.

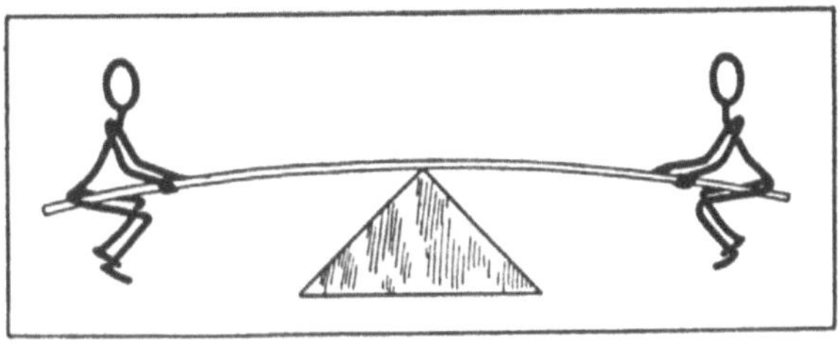

Fig. 22. Bending strength.

Bending Strength. Bending strength is that property of the metal which resists bending forces.

Torsional Strength. Torsional strength is that property of the metal which resists twisting or forces which tend to twist the metal.

Fatigue Resistance. Fatigue resistance is that property of the metal which resists changes in the crystal structure of the metal which would lead to breaking under continued vibration, hammering, or various repeated stresses.

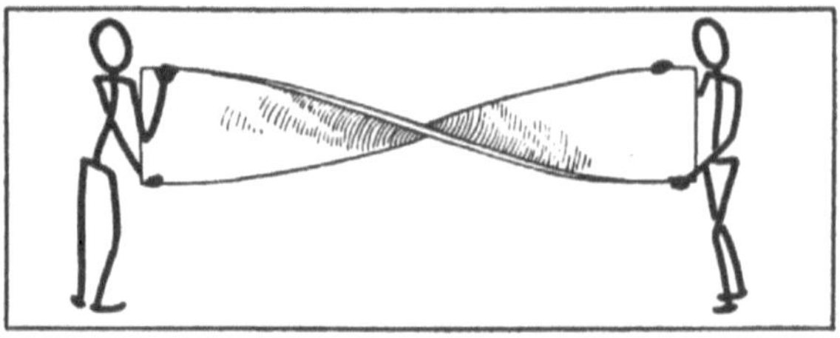

Fig. 23. Torsional strength.

Strength-Weight Ratio. Strength-weight ratio is the strength of the metal compared with its weight. For example, some of the aluminum alloys are as strong as some of the steels and weigh approximately ⅓ as much.

Fig. 24. Work-hardening.

Work-Hardening. Work-hardening is used to describe the fact that most metals become harder when worked cold, such as being rolled into sheets or rods, or hammered into shape.

III ALUMINUM AND ALUMINUM ALLOYS

Aluminum is a comparatively new metal in industry. It is said that the first article made of aluminum appeared in 1855. It is only within the present century that extensive use has been made of aluminum and its alloys. About 1909 the discovery was made that if aluminum was alloyed with small percentages of other metals, an alloy was produced which, when properly heat-treated, is equal approximately to mild steel in strength. This alloy weighs only about ⅓ as much as mild steel. Today, aluminum is used throughout the entire metal industry. The modern airplane depends largely upon aluminum to reduce its weight. New discoveries in the use of aluminum and its alloys are being made constantly.

Aluminum is obtained commercially from the mineral, bauxite. Bauxite is an impure oxide of aluminum. The earth's surface contains tremendous amounts of aluminum in the form of aluminum compounds. It is estimated that approximately 8 per cent of the earth's surface is composed of aluminum. Aluminum oxides are found in practically all clays, soils, and rocks. Bauxite is the principal mineral from which aluminum may be extracted commercially. The cost of extracting aluminum from common clay would be too great at the present time to be profitable.

The first step in obtaining aluminum from bauxite is to obtain a pure aluminum oxide. The pure aluminum oxide is mixed with the mineral cryolite, which is a compound of sodium, aluminum, and fluorine. This mixture is melted, and the pure aluminum is obtained by passing an electric current through the molten mixture. The aluminum obtained in this way is usually remelted to purify it for use in industry.

Aluminum has a comparatively low melting point and shows no color changes during heating. The metal will collapse suddenly and become liquid apparently all at once when the melting point is reached.

Aluminum is said to be "hot short." Hot short is a term used in connection with metal when it loses its ductile properties at temperatures near its melting point. Ductility is that property which tends to hold the metal together and allows it to be pulled or stretched without breaking. Aluminum melts at 1220° F. and, at 930° F., loses 92 per cent of its original strength. The heat conductivity of aluminum is high. Alu-

Fig. 25. Pouring molten aluminum. (Courtesy Reynolds Metals Company)

minum oxidizes readily when exposed to the air. A thin layer of the oxide, however, protects the metal from further oxidation.

Cast aluminum is formed by pouring the molten metal into molds and allowing it to cool. Much aluminum is prepared for use in industry by rolling or extrusion. Aluminum is also forged, which means that it is hammered into shape.

Fig. 26. Melting ingots of aluminum. (Courtesy Reynolds Metals Company)

Cast aluminum is brittle and is more easily worked while hot. Aluminum is usually rolled or extruded at temperatures ranging from about 600° F. to 900° F. Working at these temperatures is called "hot-working." When metals are worked at room temperature, they are said to be "cold-worked." Metals are usually hot-worked at a temperature which is above the softening or recrystallization temperature. The recrystallization temperature is the temperature at which the material tends to form crystals when cooling. The crystalline structure of cast aluminum is changed by working. The tiny crystals or grains are changed both in shape and arrangement. Aluminum is hardened by cold-working, but may be resoftened by annealing.

Annealing is the process of heating metal to a temperature above its recrystallization point and then allowing it to cool slowly. Annealing relieves strains within the metal and allows the crystals to return to their original form and shape.

The commercial forms of aluminum and its alloys are usually designated by numbers and letters. The commercial form of pure aluminum is designated by 2S and is approximately 99 per cent aluminum. Iron and silicon are common impurities in commercial aluminum. The structural strength of 2S aluminum is approximately ⅕ that of mild steel.

Copper, manganese, magnesium, zinc, nickel, and silicon are the common elements added to aluminum to make the commercial alloys.

Fig. 27. Hot-rolling aluminum sheet. (Courtesy Reynolds Metals Company)

Aluminum to which these elements have been added shows a decided increase in hardness and strength. Commercial metals are usually considered as pure when other elements are not present in sufficient amounts to change the properties of the metal.

Aluminum alloys, like the alloys of iron, when examined under a microscope, are found to contain crystals of the alloying substances, crystals of pure aluminum, and crystals made up of a combination of aluminum and the alloying substances. There are also true alloys formed wherein the alloying substances are dissolved in the aluminum.

"Strong alloys" of aluminum used for aircraft structures depend mainly on the amount of the alloying substance which is combined with the aluminum. Many of the alloys of aluminum depend for their strength and hardness upon proper heat treatment.

Annealing is commonly used to soften aluminum alloys. After aluminum has been cold-worked, it is usually heat-treated before being placed in the structure.

The annealing temperature of the following alloys is approximately 650° F.: 2S, 14S, 17S, 24S, 25S, 51S, and 61S. Commercial aluminum 3S is annealed at approximately 750° F. The aluminum alloy being

Fig. 28. Remelting aluminum scrap. (Courtesy Reynolds Metals Company)

annealed should be left at the annealing temperature until thoroughly heated throughout. The time required for proper annealing varies with the treatment the alloy has received before annealing. Some alloys should be brought to the annealing temperature slowly, and some rapidly. The rate at which the alloy is brought to the annealing temperature and the rate of cooling affect the grain size of the finished product. This, in turn, influences the hardness and strength of the alloy. As a general rule, the slower the cooling, the larger the grain or crystal size.

The alloy 17S, which has been given the name, Duralumin, commonly called "Dural," was discovered about 1909. This alloy showed a great increase in strength and hardness when heated to a predetermined temperature and then cooled rapidly by plunging into cold water.

About 4 per cent of copper dissolves in aluminum at high temperatures, while less than ½ per cent dissolves at room temperature. Sudden quenching by plunging into cold water prevents the separation of copper from the aluminum, thereby increasing the strength and hardness of the alloy. Copper seems to form a chemical compound with aluminum

Fig. 29. Shearing aluminum sheet. (Courtesy Douglas Aircraft Company, Inc.)

at comparatively high temperatures. The aluminum alloys of the Dural type are heated to the proper temperature and held at that temperature long enough to allow the complete dissolving of the alloyed metals in the aluminum. They are then quickly cooled by the use of cold water. Some of these alloys, particularly 17S and 24S, require from four to six days at room temperature to obtain their maximum hardness. Others, such as 14S and 25S, would require weeks at room temperature to reach their maximum hardness. To shorten the time of hardening, these alloys are heated to approximately 300° F. and reach their maximum hardness after several hours at this temperature. This hardening of the alloy, either at room temperature or a comparatively low temperature, is called "age hardening" or "aging."

The temperatures used in the heat treatment of 14S, A–17S, and 17S, are approximately 950° F. The alloys, 25S, 51S, 53S, and 61S, require a somewhat higher temperature of about 975° F. The length of time at which the alloy is held at the annealing temperature depends upon the thickness of the metal and the treatment to which it has been

Fig. 30. Sheet aluminum being stamped with alloy and temper designation. (Courtesy Reynolds Metals Company)

subjected before annealing. It is important that the cooling be as rapid as possible after the alloy is removed from the furnace or solution.

Material that has been heat-treated can be softened by annealing, but it is necessary to reheat the alloy to approximately 800° F. After being heated to this temperature, the material should be allowed to cool in the furnace at a rate not greater than about 50° F. per hr., until the temperature has been reduced to about 500° F. After cooling to 500° F., the alloy may be removed from the furnace and allowed to air-cool to room temperature.

Temper Designations. 2S, 3S, and 52S aluminum alloys are termed common alloys, while 14S, A–17S, 17S, 24S, 51S, and 61S are termed strong alloys. All of these alloys can be produced in the softest possible state when annealed. The S designates a wrought alloy, one that is produced by working, rolling, or extruding. Alloys that have been annealed are designated by an O; for example, 2S–O and 24S–O. 2S, 3S, and 52S alloys when produced as hard as possible are designated by an H; for example, 2S–H, 3S–H, and 52S–H. These alloys are also produced in tempers in between soft and hard, such as quarter-hard (2S–1/4H) and half-hard (2S–1/2H).

The strong alloys can be heat-treated to produce much harder and stronger material. 17S and 24S alloys are designated 17S–T and 24S–T when heat-treated. 53S and 61S can be produced in two heat-treated conditions, designated W and T, such as 53S–W and 53S–T. The W condition is not as hard as the T condition. The T condition is produced by heating material in the W condition to temperatures ranging from 320° F. to 350° F. This second heating operation is termed aging or precipitation heat treatment. When the alloy is heated and quenched to produce the W condition, it is called solution heat treatment. The fabricated condition is designated by an F; for example, 2S–F. The F is used to designate material in the "as extruded," "as rolled," or "as drawn" state. Heat-treated alloys may be further strengthened and hardened after heat treatment by rolling. For example, 17S–T after rolling would have the designation 17S–RT.

Among the new alloys of aluminum being produced at the present time are 75S, 301, and 303. These alloys show the highest tensile strength that is available today in any aluminum alloy. Some of the standard methods for working the more common alloys of aluminum must be modified when working the new alloys. These alloys may be obtained in various temper designations just as are other alloys of aluminum.

The following gives a summary of the conditions, or tempers, in approximate order of increasing hardness or strength:

O	Annealed (soft)
F	Fabricated (as extruded, rolled, or drawn)
¼H	One-quarter hard
½H	One-half hard
¾H	Three-quarters hard
H	Full hard
W	Solution heat-treated (but not aged)
T	Solution heat-treated and precipitation heat-treated (aged)
R	Rolled after heat treatment

Table I gives the general uses of the more common aluminum alloys, their mechanical properties, and their chemical composition.

Most of the aluminum alloys may be obtained in the form of sheets or plates, which are formed by passing the metal between steel rollers.

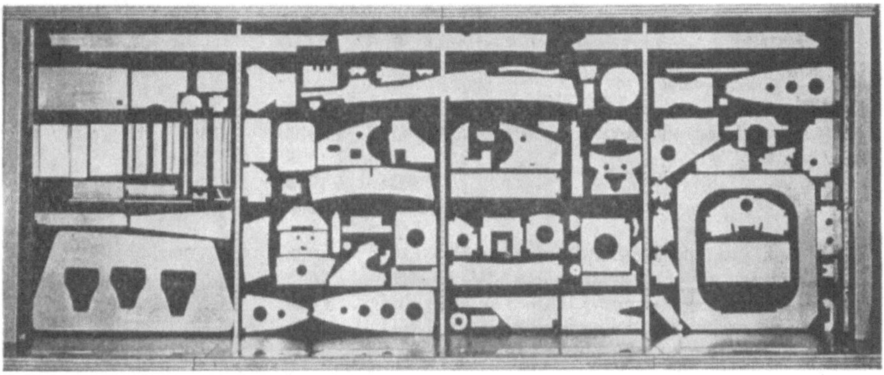

Fig. 31. Flat airplane parts prefabricated by the aluminum manufacturer. (Courtesy Reynolds Metals Company)

Most rolling is done at comparatively high temperatures, although some sheets and plates are cold-rolled.

Many airplane parts are formed by the aluminum manufacturer and shipped to aircraft manufacturers ready for installation. This procedure

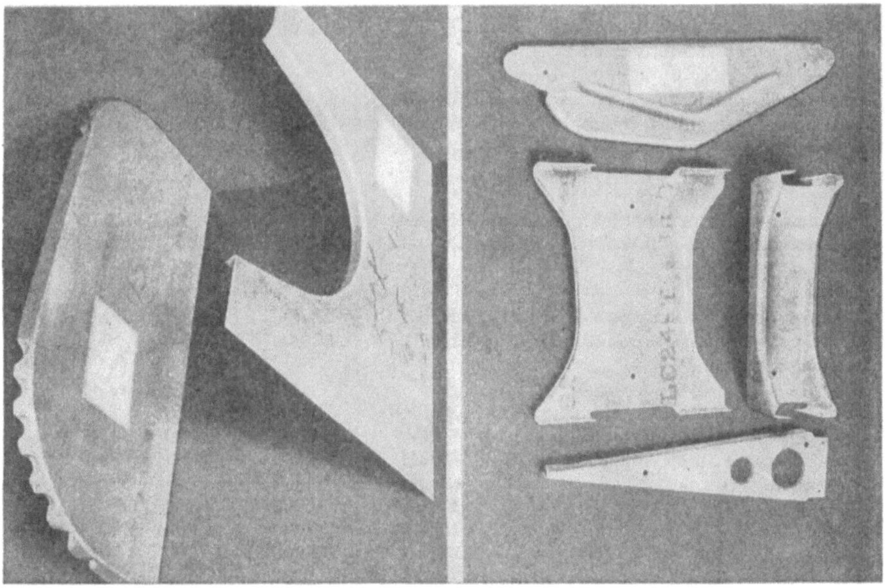

Fig. 32. Aluminum alloy parts formed on a hydraulic press using the rubber pad method. (Courtesy Aluminum Company of America)

TABLE I. ALUMINUM ALLOYS

APPLICATION	ALLOY AND TEMPER	NOMINAL MECHANICAL PROPERTIES *Yield Strength Lb. per Sq. In.*	NOMINAL MECHANICAL PROPERTIES *Tensile Strength Lb. per Sq. In.*	NOMINAL CHEMICAL COMPOSITION *(Per Cent)*
Non-Heat-Treatable-Group:				
Engineering applications involving deep drawing, welding, and aircraft non-structural parts.	2S–O 2S–¼H 2S–½H 2S–¾H 2S–H	5,000 13,000 14,000 17,000 21,000	13,000 15,000 17,000 20,000 24,000	99.0% Al (min.)
Fuel tanks and applications similar to those shown under 2S where slightly higher strength is required. If sheet or strip in relatively thin gauges and narrow widths is to be used for severe forming operations, the use of 2S is recommended.	3S–O 3S–¼H 3S–½H 3S–¾H 3S–H	6,000 15,000 18,000 21,000 25,000	16,000 18,000 21,000 25,000 29,000	1.2% Mn
Pressure tubes and vessels, fan blades, tanks, fittings, and other engineering applications involving considerable strength	52S–O 52S–¼H 52S–½H 52S–¾H 52S–H	14,000 26,000 29,000 34,000 36,000	29,000 34,000 37,000 39,000 41,000	2.5% Mg 0.25 Cr
Heat-Treatable Group:				
Rivets, hydraulic and structural fittings, and machined parts. R–317 combines excellent machinability with high strength and corrosion resistance.	17S–T R317–T	40,000	62,000	4.0% Cu 0.5 Mn 0.5 Mg
Structural applications not requiring maximum strengths attainable with Reynolds new high-strength alloys. 24S has been widely used in the past.	24S–T Pure-clad 24S–T	45,000 41,000	68,000 62,000	4.5% Cu 0.6 Mn 1.5 Mg
Applications requiring high strength, corrosion resistance, and excellent formability in W temper.	R301–W R301–T	41,000 60,000	62,000 68,000	Clad Dural Type
Requirements involving good corrosion resistance and highest strength in heat-treated condition.	R303-T Clad	71,000 67,000	77,000 72,000	Zn–Mg Type
Applications requiring good combinations of formability, weldability, corrosion resistance, and strength after heat treatment.	53S–O 53S–W 53S–T	7,000 20,000 33,000	16,000 33,000 39,000	0.7% Si 1.3 Mg 0.25 Cr
	61S–O 61S–W 61S–T	8,000 21,000 39,000	18,000 35,000 45,000	0.25% Cu 0.6 Si 1.0 Mg 0.25 Cr

APPLICATION	ALLOY AND TEMPER	NOMINAL MECHANICAL PROPERTIES		NOMINAL CHEMICAL COMPOSITION (*Per Cent*)
		Yield Strength Lb. per Sq. In.	*Tensile Strength Lb. per Sq. In.*	
Alloys-For-Forging Group:				
Structural fittings and other forged parts requiring high strength and hardness.	14S	50,000	65,000	4.4% Cu 0.8 Si 0.8 Mn 0.4 Mg
Aircraft and engine parts requiring good strength and corrosion resistance.	17S	30,000	55,000	4.0% Cu 0.5 Mn 0.5 Mg
Engine pistons and parts requiring good strength and hardness at elevated temperatures.	18S	40,000	55,000	4.0% Cu 0.5 Mg 2.0 Ni
Propeller blades and applications requiring good forging qualities and relatively high strength.	25S	30,000	55,000	4.5% Cu 0.8 Si 0.8 Mn
Applications similar to those given under 18S.	32S	40,000	52,000	0.9% Cu 12.5 Si 1.0 Mg 0.9 Ni
Use for intricate forgings requiring good corrosion resistance.	A51S	34,000	44,000	1.0% Si 0.6 Mg 0.25 Cr

(Courtesy Reynolds Metals Company)

brings about a great saving in shipping and storage space and relieves the aircraft manufacturer of the cost of handling waste material and scrap. The airplane manufacturer does not have to install machinery for forming the various parts.

Aluminum may be furnished in the form of rods or wire. The material is designated as rods when the diameter is more than 3/8 in. When the diameter is less than 3/8 in., it is called wire. Aluminum rods and wire may be formed by rolling, extruding, or drawing. Many aluminum parts are formed by forging.

Strong aluminum alloys are extruded in a number of different shapes for aircraft structures. The metal formed by extrusion is squeezed through openings in steel plates much as toothpaste is squeezed out of a tube. The shape of the extruded metal varies, depending upon the shape of the hole through which it is squeezed. Aluminum is extruded at temperatures ranging from 600° F. to 1000° F. The extruded material is usually annealed, heat-treated, straightened, and sawed into the proper length before it is ready for use.

Drawing is the process by which metal is pulled through an opening and it is often performed while the metal is at room temperature. This

Fig. 33. Airplane parts prefabricated by the aluminum manufacturer. (Courtesy Reynolds Metals Company)

is called "cold-drawing" or "cold-finishing." Aluminum and aluminum alloys may be obtained in the form of tubing, which may be either extruded or drawn.

Much of the aluminum alloy sheet used in aircraft fabrication is alclad. Alclad is a sheet of alloy covered on both sides with a sheet of pure aluminum. These sheets of pure aluminum are approximately 5 per cent of the total thickness of the sheet. Alclad is formed by placing a sheet of pure aluminum on each side of the ingot of aluminum alloy before rolling. The pressure of the rolls forms a firm bond between the pure aluminum and the alloy. The pure aluminum protects the alloy from corrosion.

Army-Navy Standards. There are several standards recognized in the aircraft industry. AN, Army-Navy Standards; AC, Air Corps Standards;

Fig. 34. Extruded shapes of aluminum alloy. (Courtesy Aluminum Company of America)

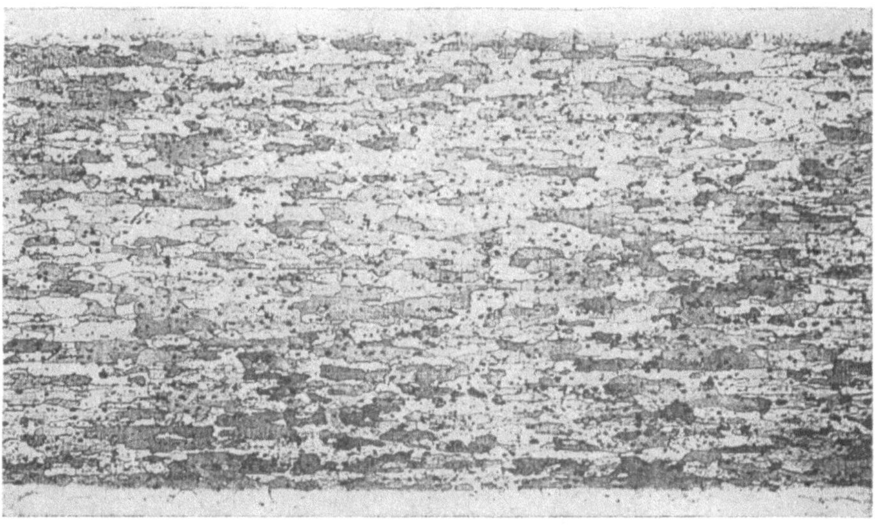

Fig. 35. The cross section of a piece of .040 inch alclad 24S–T sheet. Note the close bond between the pure aluminum and the aluminum alloy. (Courtesy Aluminum Company of America)

LS, Lockheed Standards; BAC, Boeing Aircraft Standards. "Standards" stand for approved specifications of material, dimensions, threads, shape, etc., of parts such as nuts, bolts, washers, screws, clips, fittings, material, etc. In order that the workman or mechanic may obtain the exact part or material required, all he has to do is call for it by the number designating the code used. For example, AN 4–14 means a bolt ¼ in. in diameter, 1½ in. long, with a hexhead measuring 7/16 in. from flat to flat and 5/32 in. thick with a 1/32-in. collar, made of nickel steel, and a ¼-in. 28-thread specification AN–66G–S–136, with a 1 1/16-in. grip and drilled for a cotter key size No. 48 and cadmium plated. By use of code numbers for all parts and materials, the long, complicated description of these parts is avoided when calling at the supply room for material.

IV MAGNESIUM AND MAGNESIUM ALLOYS

Because of its lightness, magnesium is coming into wide use in aircraft structures. Its low specific gravity of 1.74 makes it the lightest of the structural metals. Magnesium is a silvery white metal which, when alloyed with aluminum, zinc, manganese, and other metals, produces an alloy which has the highest strength-weight ratio of any of the commonly used structural metals. This metal weighs only about one-fifth as much as steel and about two-thirds as much as aluminum. The melting point is about 1204° F. Magnesium tends to combine readily with oxygen, and pure magnesium will burn violently. It may be worked readily by machining and be formed by standard methods. Magnesium does not form in the cold condition as readily as it does at temperatures ranging from 250° F. to 600° F.

The production of magnesium was started by the Dow Chemical Company in 1915. The alloys of magnesium compare favorably with other structural metals in corrosion-resistant qualities. Magnesium alloys may be obtained in the form of castings, extrusions, forgings, and sheet. They harden rapidly when cold-worked, but at temperatures approximating 400° F. may be worked without hardening. Because of electrolytic action, magnesium has a tendency to corrode when in contact with other metals. When it is necessary that magnesium be in contact with other metals, insulating material of some type must be used in the joint.

Magnesium alloys may be fastened by riveting, spot welding, and electric or gas welding. It is necessary, when welding magnesium by either the electric arc or the gas flame, to protect the hot metal from the oxygen of the air either by a noninflammable gas, such as helium, or by a suitable flux which will keep the metal covered.

Note: All figures and tables in Chapter IV are used by the courtesy of the Dow Chemical Company.

Table II gives the typical analysis of commercially pure magnesium.

TABLE II. TYPICAL ANALYSIS OF COMMERCIAL MAGNESIUM

ELEMENT	PER CENT
Aluminum	0.003
Copper	0.003
Iron	0.03
Manganese	0.08
Nickel	0.001
Silicon	0.005
Magnesium (by difference)	99.878

Table III shows the composition and use of magnesium alloys with their letter designation.

TABLE III. COMPOSITIONS AND USES OF MAGNESIUM ALLOYS

DOW-METAL ALLOY	NOMINAL COMPOSITION * (*Per Cent*)			USE
	Al	Mn	Zn	
C	9.0	0.1	2.0	Sand and permanent mold castings. Heat-treatable.
FS–1	3.0	0.3	1.0	Extrusions, plate, sheet, and strip.
G	10.0	0.1	—	Permanent mold castings. Heat-treatable.
H	6.0	0.2	3.0	Sand castings. Heat-treatable.
J–1	6.5	0.2	1.0	Extrusions, forgings, and sheet.
M	—	1.5	—	Extrusions, forgings, plate, sheet, and strip. Used also for special sand castings.
O–1	8.5	0.2	0.5	Extrusions and forgings. Heat-treatable.
R	9.0	0.2	0.6	Die castings.

* Remainder magnesium.

Magnesium alloys do not corrode in dry air. Ordinary air, such as is encountered in inland districts, does not cause excessive corrosion. The metal does become covered with a thin gray film consisting of the oxide and carbonate of magnesium. The presence of salt in the air, as encountered along the sea coast, or salt water increases corrosion. Magnesium not only corrodes in contact with other metals, but will corrode when in contact with wood in the presence of moisture. Any particle of foreign material which may be embedded in the surface of magnesium alloys during the process of working will cause corrosion. These tiny particles act as minute electric cells. Magnesium alloys should not be allowed to rub against steel, be cleaned with a wire brush, steel wool, or emery cloth, nor be sandblasted with any material. The metal in the area of a gas weld must be thoroughly cleaned of all flux or impurities to prevent corrosion. Magnesium alloys exposed to corrosive conditions should be protected by one or more coats of zinc chromate primer covered with suitable coats of paint.

Standard rivet heads and riveting practices may be used in fastening magnesium alloys. The rivets commonly used are of the following aluminum alloys: 2S, 3S, 53S–T61, 56S–O, 56S–¼H, and A17S–T. The

Fig. 36. Rolling magnesium alloy sheet.

alloy, 56S, is recommended because there is less possibility of corrosion. Rivets of steel, copper, brass, and other heavy metals should not be used because of the danger of electrolytic corrosion. Magnesium alloys are not used for rivets because of the rapid work-hardening which takes place when a rivet is driven cold. Rivets of aluminum alloys other than 56S should be dipped in a zinc chromate primer to reduce the possibility of corrosion. All riveted assemblies of magnesium must be painted to protect them from corrosion. If it is necessary in an emergency, such as a field repair, to use rivets of steel, copper, or brass, the rivet should be chromium plated and dipped in zinc chromate primer before driving. When sheets of magnesium alloys are riveted to dissimilar metals, the metals should be separated by a gasket of corrosion-resistant material or a suitable sealing compound. The joint between magnesium sheets which are to be riveted should be painted with two coats of zinc chromate primer before assembly. It is advisable to coat all rivets, even the alloy 56S, with zinc chromate primer before driving.

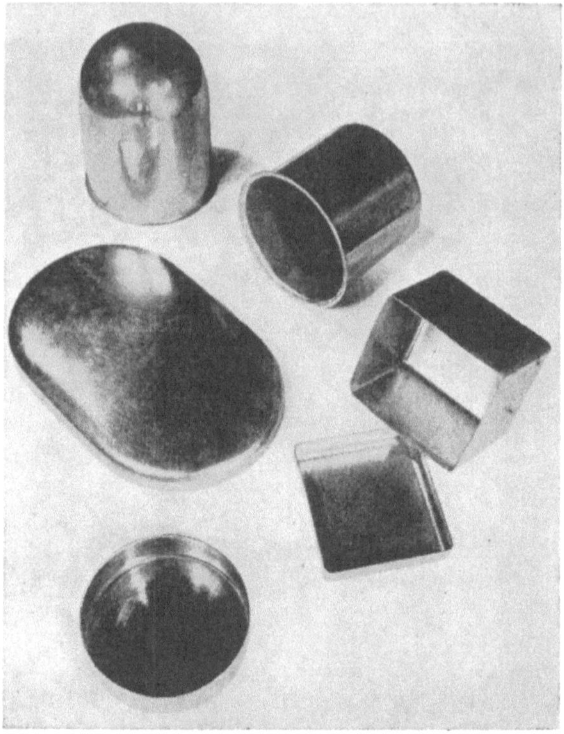

Fig. 37. Drawn parts formed from magnesium sheet.

After forming parts from magnesium alloys, the parts should be submerged in a solvent to remove all impurities from the surface of the metal. The first bath should consist of an alkaline solution which will remove all traces of grease and oil. The material should then be treated to an anodizing solution, such as a chromic acid bath, rinsed in cold running water, and protective primers and paint applied.

V IRON AND IRON ALLOYS

Without the alloys of the various metals, modern industry could not exist. The automobile contains approximately 70 different alloys of iron alone. The airplane of today could not be built without making use of the alloys of iron and aluminum.

Only small quantities of pure iron have been found in nature. Iron combines readily with the oxygen of the air and is almost always found in the form of iron oxide. Iron oxide forms the common ore from which the iron of commerce is obtained.

Common red rust is an oxide of iron. Most of the iron ores are the red oxide of the metal. The iron is obtained from its ores by a smelting process carried out in huge furnaces. The iron ore is mixed with coke, which is used as a fuel to furnish the heat, and placed in the furnace with limestone. The coke also furnishes carbon which combines with the oxygen in the ore, forming the carbon dioxide. This sets free the iron from its oxide. The limestone acts as a flux.

A flux is a substance which combines readily with the impurities in the mixture. The flux also lowers the melting point of the mixture and forms the slag which carries off the impurities, allowing the metal to be drawn off from the bottom of the furnace in a more or less pure state. The iron which is drawn off from the furnace is cast in sand molds. The ingots, which weigh approximately 100 lb. each, form the pig iron of commerce. The pig iron contains a high percentage of carbon from the coke used in the furnace, as well as other impurities, and is not suitable for most commercial use. The pig iron must be refined by resmelting, which removes the objectionable impurities and reduces the carbon content.

Steel is iron which contains a varying percentage of carbon. When the carbon content is less than $\frac{1}{10}$ of 1 per cent, the product is considered to be iron. Most so-called iron articles are not iron but are alloys of iron.

The following table is a general classification of the properties of steel, based on the amount of carbon present:

	PER CENT
Dead soft steel	0.10 maximum carbon
Soft steel	0.20 maximum carbon
Mild steel	0.15 to 0.25 carbon
Medium steel	0.25 to 0.45 carbon
Hard steel	0.45 to 0.85 carbon
Spring steel	0.85 to 1.15 carbon

The American Iron and Steel Institute has adopted the following definition for carbon steels: "Steel is classed as carbon steel when no minimum content is specified or guaranteed for aluminum, chromium, columbium, molybdenum, nickel, titanium, tungsten, vanadium, or zirconium; and when the minimum content of copper which is specified or guaranteed does not exceed 0.40%; or when the maximum content which is specified or guaranteed for any of the following elements does not exceed the respective percentages hereinafter stated: manganese, 1.65%; silicon, 0.60%; copper, 0.60%."

There are many special trade names for steel, such as tool steel, spring steel, firebox steel, high-speed steel, boiler steel, and automobile steel.

Wrought iron is an alloy of iron and carbon in which the carbon content is less than 0.10 per cent, and the manganese content is less than 0.07 per cent. Wrought iron is tough, ductile, and soft, and is easily worked.

When iron has a carbon content of less than 0.06 per cent, it is called low-carbon iron. Low-carbon steels, or mild steels, have a range of carbon content from 1/10 to 3/10 of 1 per cent. Low-carbon steels are easily worked.

Medium-carbon steels are somewhat more difficult to work. They are harder than the low-carbon steels. Tool steel and high-speed steels, which have a carbon content as high as 1.5 per cent, are difficult to work and require special heat treatment. These steels are not usually furnished in sheet form.

Some of the more common alloys of steel are molybdenum steel, chromium steel, chromium molybdenum steel, chromium vanadium steel, manganese steel, manganese vanadium steel, nickel steel, nickel chromium steel, chromium molybdenum vanadium steel, copper nickel steel, and molybdenum nickel steel. Most of these alloys have trade names and are used widely throughout the aircraft industry.

Chromium molybdenum steel is used extensively in aircraft manufacturing and does not require heat treatment.

Stainless steels have come into use in the last few years. Stainless steels are alloys of iron, chromium, carbon, nickel, and other metals. Stainless steels are known to the trade by their number classification, such as 18–8 and 25–20. One of the most common stainless steels is 18–8, which contains approximately 18 per cent chromium, 8 per cent nickel, and from 0.08 per cent to 0.20 per cent carbon alloyed with iron.

There are a number of "clad" steels in use such as copper-clad, nickel-clad and stainless-clad. Clad steels are built up by covering a sheet of the metal on one or both sides with a sheet of other material. These outer layers are usually used to prevent corrosion, although they may add strength and wear-resisting qualities. Galvanized steels are formed by dipping sheets of steel into molten zinc.

Terneplate is sheet iron or steel which has been coated with an alloy of lead and tin. It is commonly used to make patterns or templates in sheet-metal work.

Sheets coated with tin are called tin plate; with nickel, nickel plate; and with chromium, chromium plate.

Fig. 38. Diagrammatic drawing to show grains of iron alloys in high carbon steel. The solid black shows a particle of foreign matter.

The properties of steel are affected by both its physical and chemical composition. The heat treatment which steel receives affects both its physical and chemical properties. The following are some of the substances present in steel: ferrite, cementite, austenite, pearlite, primary troostite, secondary troostite, martensite, and sorbite. These materials, the physical constituents of steel, may be observed and identified through microscopic observation. The material to be examined must be carefully polished and etched with an acid solution which corrodes the metal and brings out the grain structure. This corrosion or etching indicates the grain or crystal boundaries and assists in their identification. Etching will also show the presence of nonmetallic impurities.

The temperature at which steel is suddenly cooled modifies its physical properties. When steel is heated at a known rate and the heating curve is recorded, the curve is not smooth and regular, but shows variations. Any variation in a heating or cooling curve indicates changes in the composition of the substance, which will affect its physical properties.

(See Figure 39.) The temperature between the various points where changes take place is known as the "critical range." These irregularities show that heat is absorbed or given off at these particular temperature points. When steel is heated to a point where the cooling or heating curve varies from normal, the various constituents of the metal undergo definite changes. Substances may be dissolved, but upon cooling, they may separate and form crystals. The amount of separation is controlled by the rate of cooling. The separation may be stopped by sudden cooling at any point.

Ferrite is practically pure iron, but may contain small quantities of other substances. Ferrite is present in all steels containing less than

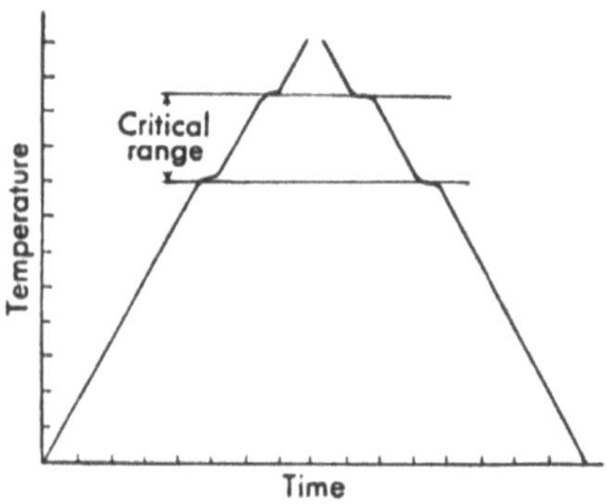

Fig. 39. Heating and cooling curve to show how the critical range might appear on the curve.

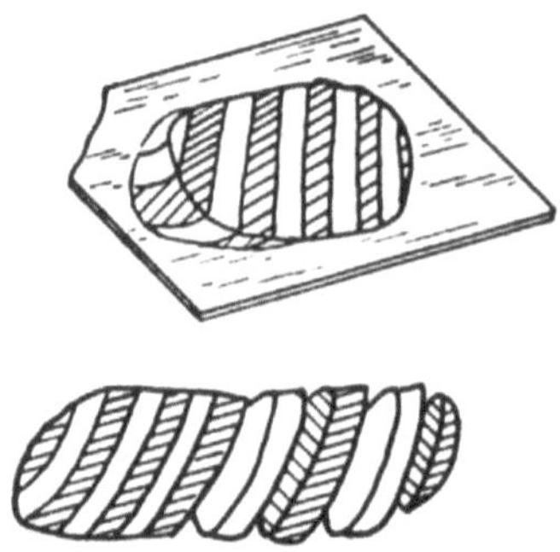

Fig. 40. A greatly magnified drawing to show the structure of a pearlite grain.

0.83 per cent carbon. When steel containing less than 0.83 per cent carbon is allowed to cool slowly from a critical temperature, it is comparatively soft, ductile, and easily worked. It has a hardness of approximately 90 Brinell, and a tensile strength of about 40,000 lb. per sq. in. Ferrite is strongly magnetic.

Cementite is a chemical compound of iron and carbon containing 93.33 per cent iron and 6.67 per cent carbon by weight. It has a definite chemical composition. Cementite is present in all steels containing more than 0.83 per cent carbon. It is extremely hard, having a Brinell hardness of about 650. Cementite is magnetic, but not ductile, and forms one of the hardest constituents of steel.

Austenite is a solid solution of carbon and other elements. Austenite is formed when carbon steels are heated to temperatures within or above the critical range. A maximum of 1.7 per cent of carbon will dissolve in austenite. When steel is cooled below the critical temperature, austenite breaks down and forms ferrite and pearlite, depending upon the rate of cooling and the amount of carbon present. Slow cooling causes a larger percentage of ferrite to be formed and results in a softer steel. Austenite is not magnetic, and its presence may be determined by measuring the loss of magnetism.

Martensite is produced when carbon steels are cooled rapidly. The amount of carbon present determines the hardness. When only a small amount of carbon is present, the sample must be cooled rapidly in order to retain martensite. Ordinary files represent this degree of hardness. Rapid cooling develops greater strength and hardness than slower cooling. Steel containing martensite is difficult to work.

Fig. 41. A drawing to show how grains of iron alloy, pearlite, and iron carbide may be surrounded by pure iron in a low carbon iron.

Fig. 42. A drawing to show scattered grains of pearlite among grains of pure iron in a very low carbon iron.

When steel is cooled slowly, troostite is formed. Troostite is harder and less ductile than pearlite. Troostite is less hard and more ductile than martensite.

Pearlite is formed by the alternate separation of ferrite, cementite, and austenite. When the carbon percentage is 0.83 per cent, ferrite and cementite form alternate, thin, platelike layers which, when viewed under the microscope, have a pearl-like sheen. When there is less than 0.83 per cent of carbon in the steel, the amount of ferrite is greater than the amount of pearlite formed, and there is a matrix of ferrite, binding together the grains of pearlite. When more than 0.83 per cent of carbon is present, the structure is a mixture of pearlite and cementite. When an alloy of this type is cooled slowly from the austenite state, a high degree of separation takes place. If the cooling is rapid, there is less chance for a separation of the ferrite and cementite, and a hard alloy is obtained.

Steel in the martensite phase may have internal stresses relieved and ductility increased by reheating to temperatures below the critical range. This process is known as tempering. The heating to a temperature below the critical range allows a gradual separation of cementite. When steel is tempered at a temperature between 400° and 750° F., a gradual change takes place according to the degree of heat. Due to the separation of the iron carbide, a partial separation of the martensite takes place. This structure in tempered steel is called secondary troostite. This increases the toughness, but the strength and hardness are decreased.

Steels are often classified according to the method used in their preparation.

Crucible Steel. This steel is made from low carbon iron and is soaked at melting heats in the heating furnace gases in a covered crucible.

Bessemer Steel. Bessemer steel is made by blowing air through molten pig iron while it is in a Bessemer container called a converter. The carbon and other impurities are burned out. Limestone is added during the refining of the pig iron to keep the slag basic. Phosphorus from the steel is absorbed by the slag and the lining of the converter. This process leaves the steel with a low phosphorus content.

Open Hearth Steel. Open hearth steel is made by holding the pig iron in a molten state in the furnace until the carbon and other impurities are reduced to the desired proportions. In the open hearth furnace the flame plays over the surface of the molten metal. The open hearth process can be more closely controlled than the Bessemer process and produces a very reliable steel.

Electric Furnace Steel. The electric furnace method resembles the crucible steel method except that the heat for melting and refining is supplied by electricity instead of gas, coke, or oil. The electric method and the crucible method are the most expensive methods used for producing steel in large quantities.

Cast iron is iron containing 2½ per cent, or more, carbon and is cast from the molten state. Cast iron is brittle and cannot be forged or hammered into shape.

Malleable iron is iron containing from 2 per cent to 2.75 per cent carbon and is cast from the molten state. This steel is forged at comparatively high temperatures, depending upon the amount of carbon present. Carbon, manganese, silicon, sulphur, and phosphorus are elements usually present in cast steel. Nickel, chromium, vanadium, tungsten, and molybdenum are the metallic elements most commonly used in the making of steel alloys.

The Society of Automotive Engineers has developed a numbering code for basic steels which is generally accepted throughout industry. The American Iron and Steel Institute has also developed a numbering code which is more strict concerning the limits of alloy contents and makes additional provision for denoting other limits of quantity and properties. The American Society for Testing Materials has also developed specifications for use in testing various materials.

The following is a brief explanation of the Society of Automotive Engineers' system for numbering basic steel:

No. 1. Carbon steel
No. 2. Nickel steel
No. 3. Nickel chromium steel
No. 4. Molybdenum steel
No. 5. Chromium steel
No. 6. Chromium vanadium steel
No. 7. Tungsten steel
No. 8. Silicon manganese steel

Example No. 1. S.A.E. steel No. 1045
First figure (1) carbon steel
Second figure (0) no alloys
Last two figures (45) carbon content from 0.40 per cent to 0.50 per cent

Example No. 2. S.A.E. steel No. 2340
(2) — nickel steel
(3) — 3 per cent alloy; of nickel
(40) — carbon content from 0.35 per cent to 0.45 per cent.

Example No. 3. S.A.E. steel No. 71350
(7) — tungsten steel
(13) — 13 per cent alloy; tungsten
(50) — carbon content from 0.45 per cent to 0.55 per cent.

VI PRINCIPAL SHEET METALS USED IN AIRCRAFT FABRICATION AND THEIR IDENTIFICATION

Almost all of the metals and their alloys may be obtained in the sheet form. Metals are prepared in the sheet form by passing between heavy steel rollers. The composition of the sheet may be the same as in tubes, bars, forgings, or castings.

Material up to a thickness of ⅛ in., which is approximately 11 gauge, is usually classified as sheet. When the rolled material exceeds ⅛ in. in thickness, it is usually classified as plate. Very thin material is called "foil" or "leaf." The sheets used in aircraft construction are usually designated by their thickness in decimals of an inch rather than by the gauge, which is more commonly used in other industries. The thickness of the metal in terms of gauge varies with the different numbering systems used. For example, in the American Wire Gauge and Brown and Sharpe systems, 16 gauge has a decimal equivalent of 0.051 in.; while in the Birmingham Wire Gauge and Stub systems, 16 gauge has a thickness of 0.065 in. These thicknesses are roughly equivalent to 1/16 in.

The sheet-metal worker will be required to work with many metals and their alloys and should be able to determine the kind of metal with which he is to work. He must know whether the metal to be worked has been heat-treated, cold-worked, or annealed. When possible, this information should be obtained from blueprints and specifications furnished by the manufacturer. Many times this information may be obtained from the identification markings placed upon the material itself. However, it is often necessary for the sheet-metal worker to determine the kind of material in a part to be replaced.

The following list of properties and tests will assist the sheet-metal worker in identifying most of the metals with which he will be called upon to work:

1. The general appearance of the metal.
2. The appearance of the freshly polished surface.
3. Chipping the metal with a cold chisel.
4. The spark test.
5. The action of the metal while being formed.
6. The weight of the metal.
7. The hardness of the metal.

The newly machined or filed surface should also be examined for color, smoothness, and luster.

The chip test is made by removing a small amount of material from the edge of the sample with a sharp cold chisel. The material removed will vary from small, broken fragments to a continuous strip. The chip may have smooth, sharp edges, be coarse grained or fine grained, or have sawlike edges where cut. The size of the chip is important in identifying the sample. The ease with which chipping takes place also assists in the identification.

The spark test is made by holding a sample of the material against a power grinder. The sparks given off or the lack of sparks assists in identifying the sample. The length of the spark stream, its color, and the type of sparks are items which should be noted. Figure 43 shows the various types of sparks obtained by the use of a grinding wheel.

The difference in hardness between metals and alloys can be estimated by the use of a sharp flat file. If the material is cut by the file with extreme ease and tends to clog the spaces between the file teeth, it is classified as very soft. If the material offers some resistance to the cutting action of the file and tends to clog the file teeth, it may be classified as soft. If the material offers considerable resistance to the file but can be removed by a considerable effort, the material may be classified as hard and it may or may not have been heat-treated. If the material can be removed by extreme effort and in small quantities by the file teeth, it may be classified as very hard and it has probably been heat-treated. If the file slides over the material and the file teeth are dulled, the material may be classified as extremely hard and one which has been heat-treated.

White Cast Iron. On a newly fractured surface, white cast iron shows a fine, silky, silvery white, crystalline appearance. The outside of the material shows evidence of having been in a sand mold and is a dull gray. When freshly filed, it shows a silvery white surface. Chips come off in small broken fragments. It is brittle, and the chipped surface is not

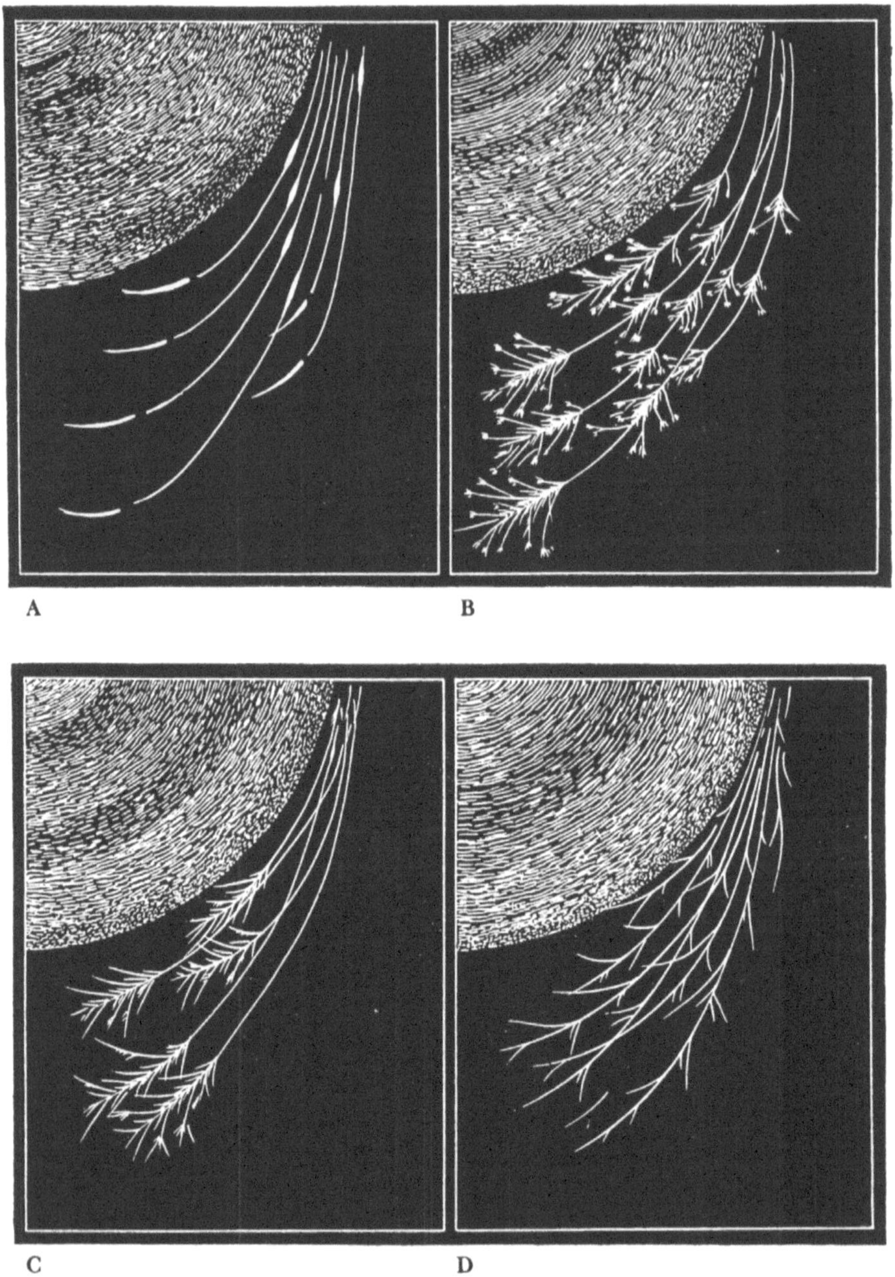

Fig. 43. Fundamental spark forms produced by holding a sample of metal against a power grinder. (A) Shafts, bud, break, arrow; (B) shafts and sprigs or sparklers; (C) shafts, forks and sprigs; (D) shafts and forks.

smooth. The length of the spark stream is approximately 20 in. The volume of sparks is small with many sparklers, which are small and repeating. The color of the spark stream close to the wheel is red, while the outer end of the stream is straw colored. White cast iron melts at a moderate rate under the blowpipe and becomes a dull red before melting. A medium film of slag develops which is quiet, tough, and can

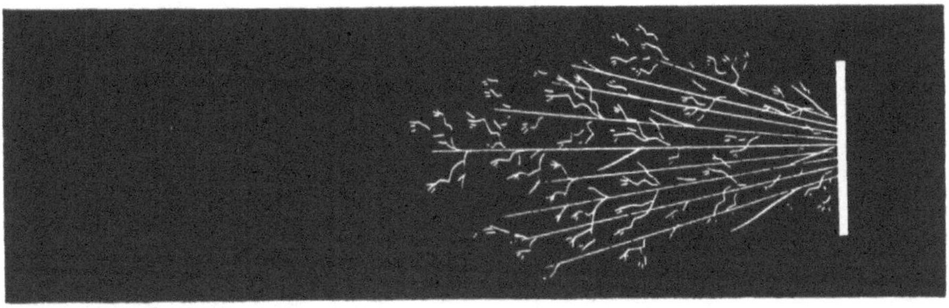

Fig. 44. Sparks from white cast iron produced by a grinding wheel. Note the many sparklers and sprigs.

be broken up. The molten puddle is fluid, watery, and reddish white and does not show sparks. The depression under the flame disappears when the flame is removed.

Gray Cast Iron. The newly fractured surface of gray cast iron is somewhat the color of white cast iron, but a little darker. The outside shows evidence of having been in a sand mold and is dull gray. A freshly filed surface is fairly smooth and light silvery gray. The chips are about

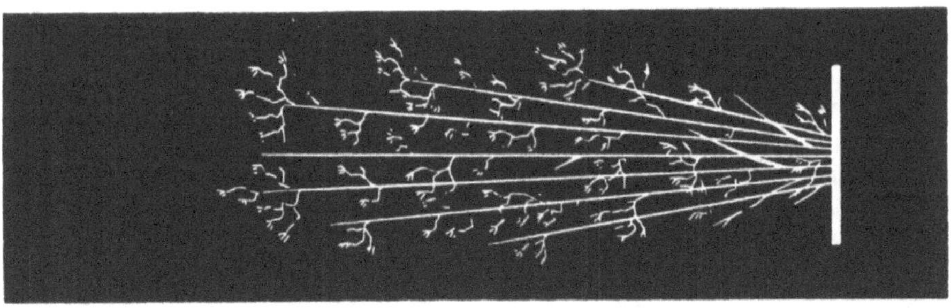

Fig. 45. Sparks produced from gray cast iron by a grinding wheel.

⅛ in. in length. It is not easily chipped, as the chips break off, preventing a smooth cut. Spark tests usually develop a spark stream about 25 in. long, small in volume with fewer sparklers than white cast iron. The sparklers are small and repeating. Part of the stream close to the grinding wheel is red, and the outer end of the stream is straw colored. Gray cast iron melts moderately fast under the blowpipe and becomes a dull red before melting. A thick film of slag forms which is quiet and tough, but is possible to break up. The molten puddle is fluid, watery, reddish white, quiet, and shows no sparks. The depression under the flame disappears when the flame is removed.

Malleable Iron. Malleable iron shows a rather fine crystalline, dark gray surface when freshly broken. There is evidence of its having been

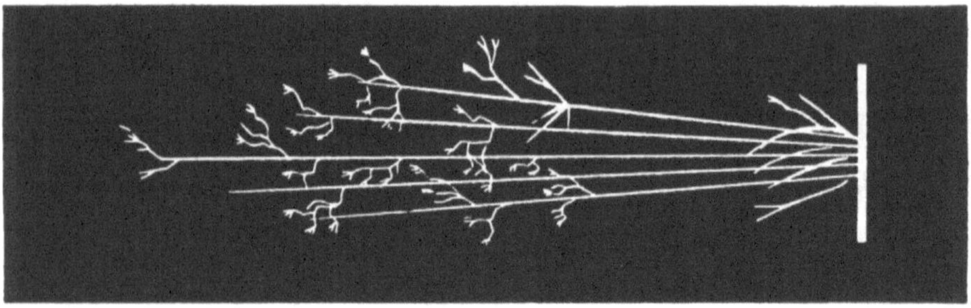

Fig. 46. Sparks produced from malleable iron by a grinding wheel. Note the many sprigs or sparklers toward the end of the stream.

in a sand mold, and the outside is dull gray. The freshly filed surface is smooth and light silvery gray. The chips are not as small as cast iron, being from ¼ to ⅜ in. long. Malleable iron is tough and hard to chip. The spark test shows a stream of sparks about 30 in. in length, of moderate volume, and with many sparklers toward the outer end of the stream, which are small and repeating. The entire stream is

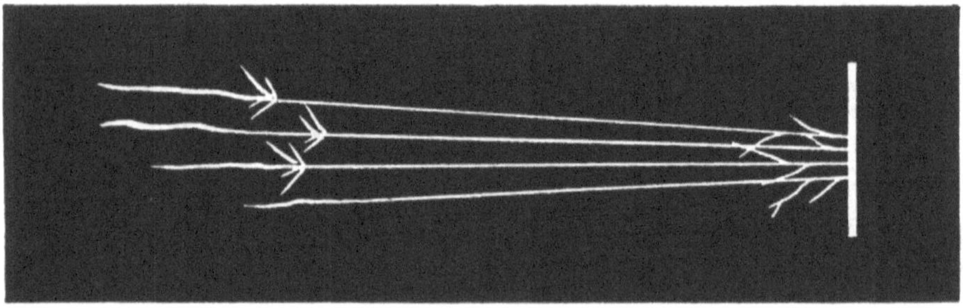

Fig. 47. Sparks produced from wrought iron by a grinding wheel. Note the forks and sprigs or sparklers toward the end of the stream.

straw colored. Malleable iron melts at a moderate rate under the blowpipe and becomes red before melting. A medium film of slag develops which is quiet and tough, but can be broken up. The molten puddle is fluid, watery, and straw colored; it boils and leaves blowholes. The surface of the puddle sparks, but the interior does not.

Wrought Iron. The surface of wrought iron, when freshly broken, is a bright gray and quite smooth. The unfinished surface is smooth and light gray, while the freshly filed surface is smooth and light silvery gray. The chips have a smooth edge and may be continuous if desired. Wrought iron is soft and easily cut or chipped. The spark stream is about 65 in. long and large in volume with few sparklers. The sparklers appear toward the end of the stream and are forked. The stream next to the grinding wheel is straw colored, while the outer end of the stream is white. Wrought iron melts rapidly under the blowpipe and becomes a bright red before melting. The slag is oily or greasy in appearance with white lines. It is quiet and easily broken up. The molten puddle is liquid and straw colored; it is not viscous and is usually quiet, although it may have a slight tendency to spark.

Low-Carbon and Cast Steel. On a freshly broken surface, low-carbon steel is bright gray with a rather fine crystalline appearance. The unfinished surface is dark gray and may have forging marks or show evidence of a mold. The newly filed surface is smooth and a bright silvery gray. The chips have smooth edges and may be continuous if desired. It is easily cut or chipped. The spark stream is about 70 in. in length, and the volume is moderately large. The few sparklers, which may occur at any place in the spark stream, are forked. The color of the spark stream is white. Low-carbon steel melts rapidly under the blowpipe and becomes bright red before melting. The slag is similar to the molten metal and is quiet. The molten puddle is liquid, straw colored, and it sparks.

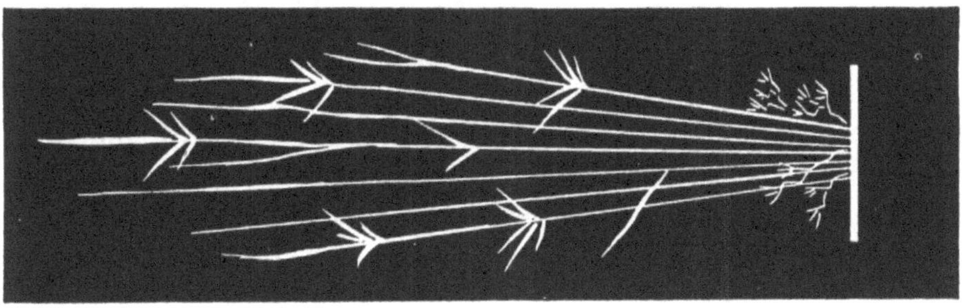

Fig. 48. Sparks produced from low-carbon and cast steel by a grinding wheel. Note the long shafts with forks.

High-Carbon Steel. On a newly broken surface, high-carbon steel is light gray and has a fine crystalline appearance. On an unfinished surface, rolling or forging lines may be noticeable. The newly filed surface is smooth and bright silvery gray. The chips show fine grain fracture, their edges are lighter in color than those of low-carbon steel, and they may be continuous if desired. The metal is hard and can be chipped. The spark stream is white, about 55 in. long, and large in volume, with many sparklers. The sparklers are small and repeating, and some of the shafts may be forked. High-carbon steel melts rapidly under the blowpipe and becomes bright red before melting. The slag is similar in appearance to the molten metal and is quiet. The molten puddle is

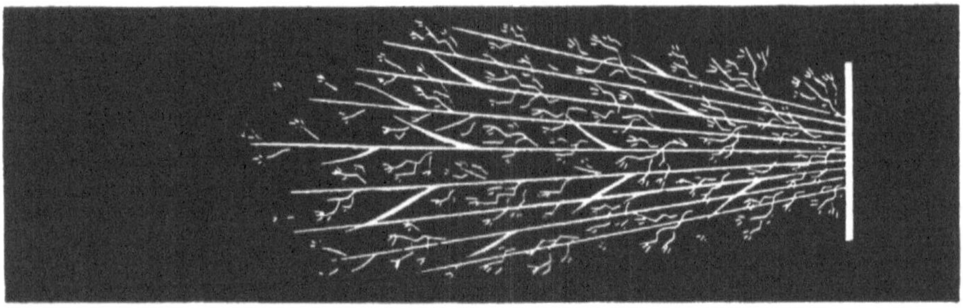

Fig. 49. Sparks produced from high-carbon steel by a grinding wheel. The stream is large in volume with many sprigs or sparklers.

lighter and sparks more freely than that of low-carbon steel. It has a cellular appearance.

Alloyed Steels. Alloyed steels vary so much in composition that the results of the various tests are not definite. The sheet-metal worker should experiment with various known alloys until he is familiar with the appearance of the various tests. For example, inconel and 18–8 stainless steel are alloys containing nickel and iron, which might be easily confused. To determine whether a sample of metal is inconel or 18–8 stainless steel, a small clean spot on the metal is treated with a few drops of solution consisting of 10 g. of cupric chloride in 100 cu. cm. of hydrochloric acid. One drop of the solution on the cleaned area should be allowed to remain in contact with the metal for two minutes. After two minutes, add with a medicine dropper, 1 drop at a time, 3 or 4 drops of water. If the sample is 18–8 stainless steel, the copper in the cupric chloride solution will be deposited on the metal, leaving a copper-colored spot. If the sample is inconel, a white spot will be left where the solution contacts the metal. Copper will not be deposited until water is added to the diluted mixture.

Stainless Steel. Stainless steel is medium gray on a newly broken surface and the grain is rather fine. The unfinished surface is usually dark gray and slightly rough. Rolling or forging lines may be noticeable. The newly filed surface is smooth and a bright silvery gray. The chips vary with the alloys. The spark test for stainless steel produces a stream approximately 50 in. long of moderate volume with a few sparklers. The sparklers are forked. The stream next to the grinding wheel is straw colored, while the outer end is white. The action under the flame varies with the composition of the alloys.

Copper. Copper is a bright red color on a newly broken surface, while an unfinished surface may be reddish brown to green, due to corrosion.

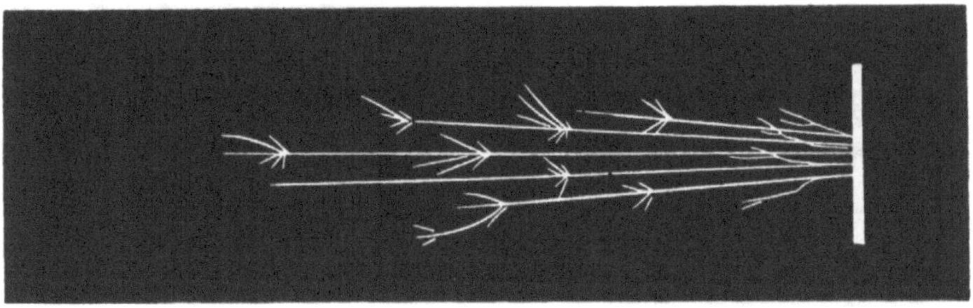

Fig. 50. Sparks produced from stainless steel. Note the long shafts, forks, and few sparklers or sprigs.

The surface is usually smooth. When freshly filed, it is a bright copper color which dulls with time. The chips are smooth, have saw edges where cut, and can be continuous if desired. Copper cuts easily and has no spark on the grinding wheel. It melts slowly under the flame and may turn black and then red. The copper color may become intense just before melting. There is little slag. The molten puddle has a mirror-like surface directly under the flame and a tendency to bubble. The puddle solidifies slowly and may sink slightly when the flame is removed.

Brass and Bronze. The color of brass and bronze may vary from red to yellow. A newly broken surface is quite rough and crystalline. An unfinished surface shows various shades of reddish yellow, yellow green or brown, and is usually smooth. A freshly filed surface varies from reddish yellow to yellow white, and is smooth and bright. The chips are smooth and have saw edges. It is difficult to obtain a continuous chip. Brass and bronze are easily cut, but are more brittle than copper. There is no spark when the grinding wheel is used. Brass and bronze melt quite rapidly under the flame and become noticeably red before melting. The slag usually appears as various quantities of white fumes during melting. The molten puddle is liquid and looks like drops of water.

Aluminum and Aluminum Alloys. Aluminum and aluminum alloys show a fine, crystalline, white surface when freshly broken. The white or light color and extremely light weight make aluminum easy to identify. Aluminum alloys are usually harder and slightly darker in color than pure aluminum. An unfinished surface may show the marks of molds or rolls and is light gray in color. A newly filed surface is smooth and white. The chips are smooth, have saw edges, and can be continuous if desired. There is no spark when aluminum is applied to the grinding wheel. Aluminum and its alloys melt faster than steel with no apparent change in color. A stiff black scum which is quiet usually forms. The

Fig. 51. Sparks produced from nickel. Note the small volume and wavy streaks.

molten puddle is the same color as the nonheated metal and is fluid. The slag is quiet. To distinguish between aluminum and aluminum alloys, it is only necessary to dip a small portion of the metal in a 10 per cent solution of caustic soda for a few seconds. The alloy will turn dark and remain dark after washing in plain water. Aluminum also turns dark, but the discoloration may be washed off with water. When the sheet is alclad, the surface will test pure aluminum, but the edge will show a dark area which is permanent. The flame test may be used to identify magnesium alloys. A small piece of the metal heated to its melting point will ignite and burn with a bright flame. Aluminum or aluminum alloys will form into a small ball of molten metal, which is characteristic of these alloys under heating.

Monel Metal. Monel metal is fairly smooth and light gray on a newly broken surface. An unfinished surface is smooth and dark gray. A freshly filed surface is smooth and light gray. The chips have smooth edges and may be continuous if desired. Monel metal chips easily. The spark stream is about 10 in. long, small in volume, and orange in color, with no sparklers. The sparks form small wavy streaks similar to nickel. Monel metal melts more slowly than steel and becomes red before melting. The slag forms a gray scum in considerable amounts and is quiet and hard to break up. The molten puddle is fluid under the slag and quiet.

Nickel. Nickel is slightly rough and almost white on a newly broken surface. An unfinished surface is smooth and dark gray. A freshly filed surface is smooth and a bright silvery white. It chips easily, and the chips have smooth edges and can be continuous if desired. The spark stream is about 10 in. long, small in volume, and orange in color. The sparks form small wavy streaks with no sparklers. Nickel melts more slowly than steel and becomes red before melting. The slag, in the form of gray scum, is quiet and hard to break up. The molten puddle is fluid under the slag and quiet.

Lead. Lead is light gray and crystalline on a newly broken surface. An unfinished surface is smooth, velvety, and from white to gray in color. A newly filed surface is smooth and white. Any shape of chip may be obtained because of the softness and it may be continuous if desired. Lead is soft enough to cut with a knife. It shows no spark on a grinding wheel and melts fast with no apparent change in color. The slag is a dull gray and is quiet. The molten puddle is white and fluid under the slag, but may boil if too hot, giving off fumes.

The aluminum alloys, 2S, 3S, and 52S, are usually used in the sheet form for nonstructural parts where high strength is not necessary, such as air scoops, cowlings, fairings, fuel and oil tanks, and parts which are to be fastened by welding. These alloys are softer than other alloys of aluminum and are more easily formed.

The strong alloys of aluminum, 17S–T, 24S–T, and 75S–T, are used in structural parts where strength is necessary, such as spars, ribs, bulkheads, stringers, and the skin. These alloys are hard, stiff, and difficult to form in their heat-treated condition. When considerable forming is to take place, the sheets of these alloys are usually formed in the annealed condition and heat-treated after forming.

Magnesium alloys are furnished to the industry under the name, Dowmetal, and contain magnesium in percentages varying from 89.9 to 98.5. The magnesium is alloyed with manganese, aluminum, silicon, zinc, and tin. These alloys are coming into use rapidly because of their high strength-weight ratio. The magnesium alloys may be used to advantage where lightness and strength are desired.

A number of alloys of steel are used extensively in the fabrication of aircraft structures. Chrome-molybdenum steel is obtainable in sheet and tube form. The sheet form is used for fittings and parts to be fastened by welding.

Stainless steels, such as 18–8, are used in the sheet form for structural members and fittings where high tensile strength and noncorrosive qualities are required. This material has recently come into considerable use as skin covering for aircraft. Stainless-steel sheets should be fastened by means of rivets or spot welding where high-strength joints are required. While stainless steel may be successfully welded with the oxy-acetylene flame, the heat of the operation tends to destroy the heat-treated qualities of the material in the vicinity of the weld.

Copper and copper alloys in the sheet form are used to some extent in aircraft fabrication where their particular corrosion-resisting qualities are desired. Sheet metal of nickel alloys, such as inconel and Monel metal, is used when heat-resistant and noncorrosive qualities are necessary, as in collector rings and exhaust stacks.

VII GLOSSARY OF TERMS USED IN AIRCRAFT SHEET METAL

aileron. A hinged or pivoted movable auxiliary surface of an airplane, usually part of the trailing edge of the wing, the purpose of which is to maintain the lateral balance of the airplane.

aircraft. Any weight-carrying device or structure designed to be supported by the air, either by buoyance or dynamic action.

airfoil. Any surface, flat or curved, rigid or movable, such as wing, aileron, or rudder, which catches the air in such a manner as to affect the motion of the plane.

airplane. A mechanically driven aircraft, heavier than air, fitted with fixed wings and supported by the dynamic action of the air.

air scoop. A scoop or hood designed to catch the air and maintain the air pressure in ballonets, internal-combustion engines, ventilators, etc.

air screw (propeller). A surface so shaped that its rotation about an axis produces a force (thrust) in the forward direction of the axis.

alclad. Sheet Dural with a thin coat of pure aluminum on both sides (aluminum-clad Dural). The purpose of the aluminum is to protect the Dural and prevent corrosion.

alignment. The state of being in line, or to draw into line.

alloy. A mixture or solid solution of two or more metals.

aluminum. Aluminum is one of the 92 chemical elements. It is the principal ingredient in all of the aluminum alloys. A coating of pure aluminum is applied to high-strength aluminum sheets to produce alclad. The chemical symbol for aluminum is Al. It has a density of 0.0943 lb. per cu. in. The melting point of aluminum is 1220° F.

aluminum alloys. Alloys formed by combining commercially pure aluminum with other metals, namely, copper, silicon, manganese, magnesium, chromium, iron, zinc, and nickel.

angle, attaching. A piece of material bent or formed into an angle which is used as a structure for attaching.

angle of incidence. *See* Angle of Wing Setting. In British terminology the angle of incidence is equivalent to the American term "angle of attack."

angle of wing setting. The acute angle between the plane of the wing chord and the longitudinal axis of the airplane. The angle is positive when the leading edge is higher than the trailing edge.

annealing. Heating of an alloy in the solid state to a temperature above the critical range or the solid solution temperature, and allowing it to cool slowly through the critical range. The term, annealing, is used rather broadly sometimes, but in general it is to (1) remove internal stresses, (2) induce softness, ductility, and toughness, (3) refine the crystalline structure.

anodize. The artificial oxidation of the surface of aluminum or magnesium alloys by means of an electro-chemical treatment.

antidrag wire. A wire designed primarily to resist forces acting parallel to the chord of the wing of an airplane and in the same direction as flight.

assembly. A group of parts fastened together to form a unit which may be installed in another assembly or in the plane itself. The assembly drawing shows dimensions needed to locate parts with respect to each other, giving instructions for drilling, reaming, trimming, etc.

bevel. Cutting or forming an angle between the edges of metal parts.

bevel, closed. The condition of any construction making an angle of less than 90° with 90° perpendicular. Thus, a closed bevel of 10° would specify that the construction formed an angle of 80° with the line perpendicular to the 90° line.

bevel, opened. Same as above except that the angle is the specified number of degrees greater than 90.

blanking, sheet-metal. The operation that shears the flat pattern (blank) from the original sheet or strip stock.

blowpipe. The unit of the welding apparatus which brings the two gases together in the correct proportions and thoroughly mixes them before they are released to form the flame.

bonding. A piece of material used to establish an electrical connection between adjoining structures or units.

brazing. Uniting of metal parts with a copper alloy which has a lower melting point than the metal being joined.

brittleness. That property of a metal which causes it to fracture when deformed.

bucker. A person holding a bucking bar.

bucking. The act of holding the bucking bar on the rivet shank to upset the rivet.

buckling. Distortion of sheet metal due to the forces of expansion and contraction caused by heat.

bulb-angle. A right angle of metal having the outer edge of one flange enlarged in a circular manner to lend additional stiffness to the part.

bulkhead. Lateral partition, complete or incomplete, of an enclosed fuselage.

burring. Removing the ragged edge around holes after drilling.

butt weld. Operation by which the edges of two surfaces are brought together, edge to edge, and are welded along the seam thus formed.

cadmium. One of the chemical elements. A silvery white crystalline metal, whose symbol is Cd. It is commonly used as a plating material (cadmium plate) because of its corrosion-resistant characteristics.

cantilever. A kind of wing design so arranged that the entire structural strength of a wing is within the wing itself, requiring no outside bracing.

carburizing flame. A gas flame having the property of introducing carbon into the metal being heated.

case hardening. The hardening of the outside of a piece of metal, leaving the inside soft and tough.

center section. Central panel of the wing structures, usually located directly over the fuselage to which the wings are attached on each side.

chord. A straight line between the leading and trailing edges of an airfoil.

chromodizing. Adding a thin film of aluminum oxide to the surface of alclad material, by immersion in a chromic acid solution.

cloth. Fabric delivered by the bleachery or finisher before it has been proofed, doped, or specially treated for aeronautic use.

cockpit. Open space in an airplane for accommodation of pilots or passengers. When completely closed, it is usually called a cabin.

compression member. A longitudinal strut between spars opposing the pull of wires in the internal drag truss of a wing.

compression rib. A rib which is built particularly strongly with either heavier cap strips or a solid web (or both) for the purpose of withstanding greater compression stress.

concentric. Circles or spheres which have a common center.

conductivity. The property by which heat travels through a metal.

conduit. A tube or trough for receiving and protecting electric wires or cables.

control surfaces. Movable airfoils by means of which the airplane is controlled in flight. They are hinged auxiliary surfaces and consist of ailerons, elevators, rudders, tabs, and flaps. The control mechanism is the means by which the control surfaces are actuated.

copper. One of the chemical elements. Its symbol is Cu. Copper is used extensively for electrical parts, being a good conductor of heat and electricity. It is used as a base metal for alloying with zinc to form brass and for alloying with tin to form bronze. Copper is quite ductile, malleable, and tenacious.

countersink. Removing material about the end of a hole in an object so that the heads of rivets, bolts, and screws may be flush with the surface of the object.

cowling. A removable covering.

datum line. A base line or reference line from which calculations or measurements are taken.

density. The weight of a substance per unit volume.

die casting. A method by which molten metal is forced into suitable permanent molds.

dies. Tools whose purpose is to impart any desired shape or impress any desired form or design on metals or materials. On those dies which shear a part out of the stock sheet, or form it from the flat sheet, the die is the female portion which the punch enters to perform the required work.

dimpling. Press countersinking with dies for flush riveting in thin material or skin under 0.050 inch thick.

distortion. The deformation of metal due to uneven heating.

drag wire. A wire designed to resist forces acting parallel to the chord of the wing and in the opposite direction of flight.

drill. An instrument for boring holes in hard substances.

durability. A general term for permanence or lastingness.

dural. The alloy of aluminum used most extensively for aircraft production. It is made by the addition of certain small percentages of copper, manganese, magnesium, and silicon — one or all of these being used to form the alloy. In all cases, the aluminum remains above 90 per cent.

edge distance. The distance from the center of the rivet to the edge of sheet or stock.

elasticity. The property of a substance to recover its original size and shape after being deformed.

elastic limit. The maximum load a metal will maintain before it begins to stretch.

electrolytic. Pertaining to electrolysis. A chemical decomposition caused by an electric current.

elongation. The amount a metal will stretch, when pulled, before it is pulled apart.

etching. The art of producing pictures or designs by means of etched plates.

exhaust-collector ring. A circular duct into which the exhaust gases from the cylinders of a radial engine are discharged.

expansion coefficient. The amount a metal will increase in size per degree rise in temperature per inch.

extrusion. The process or method by means of which the more plastic metals, such as lead, tin, aluminum, zinc, copper, etc., are made into intricate cross-sectional shapes by being forced through the proper die opening.

fairing. A member or structure for the purpose of producing a smooth outline and reducing wind resistance.

fatigue. The loss of strength in a metal due to repeated bending or application of loads.

felt. A fabric of wool (or wool and cotton, fur, or hair) made by matting the fibers together under pressure.

ferrous. Pertaining to, or derived from, iron.

ferrule. Metal fittings or wire wrappings designed to prevent splitting of wooden parts or loosening of wire terminals, etc.

final assembly. This is the assembling of component sections to make up the complete airplane.

firewall. A fire-resistant, transverse bulkhead, so set as to isolate the engine compartment from the other parts of the structure and thus confine a fire to the engine compartment as much as possible.

fitting. A general term referring to any one of the many small, and not otherwise named, parts used in fabricating the structure of an aircraft. The most important fittings are those which join two or more major assemblies, such as wing to fuselage.

flange. Any web-stiffening portion of I-beam sections, channel sections, cap strips on wing ribs or spars, etc.

flap. A hinged or pivoted airfoil, forming the rear portion of an airfoil, used to vary the effective camber.

flux. A chemical powder used to dissolve oxides, clean metal, and prevent oxidation during welding.

forging. Metallic parts fabricated by placing a preheated bar or rod under the repeated blows of a hammer and forcing the material to conform to the contours of the steel forming die.

former or false rib. An incomplete rib extending from the leading edge to the front spar to assist in maintaining the front curvature of the wing.

framework. The frame of an airplane, exclusive of the skin.

fuselage. The structure, approximately streamlined in form, to which are attached the wings and tail units of an airplane.

gauge. Any suitably shaped standard tool that has been accurately finished to some standard dimension for use in checking or measuring the finished dimension of many parts or tools or distances between various adjacent parts.

gusset. A small connection plate, usually of thin metal, to strengthen angles or joints.

hardness. The ability of a metal to resist permanent deformation.

hardness tests for metals, classifications:[1]

Brinell Hardness Test.
Rockwell Hardness Test.
Shore Scleroscope Hardness Test.

heat treatment. Changing the physical properties of metals and alloys by heating and cooling under various conditions.

[1] There are other hardness tests, but the three listed are the ones in most common use.

heat treatment of aluminum alloys. A very accurately controlled process of heating, quenching, and aging of the alloys, which develops the maximum hardness and strength characteristics of the hard aluminum alloys.

horn. A short lever attached to a control surface of an aircraft to which the operating wire or rod is connected.

hot short. The condition in which metals become extremely weak at temperatures below the melting point.

inconel. A registered trade mark of the International Nickel Company, Inc. The name, Inconel, is applied to a nickel chromium iron alloy. It contains approximately 80 per cent nickel, 14 per cent chromium, and 6 per cent iron. It has physical properties similar to stainless steel. It cannot be hardened by heat treatment, but the effects of cold-working (strain hardening) can be relieved by annealing.

inspection hole. A small hole having an easily removable cover.

iron. One of the chemical elements. Its symbol is Fe. It is used in the manufacture of steel and cast iron.

jig. Any rigid structure or mechanism, either of wood or metal, which holds parts while they are being fabricated (drilled, welded, sawed, etc.) prior to assembly, or which holds the component parts of a structure while it is being assembled or disassembled.

joint. A union of two pieces of stock.

kerf. The slot or space from which the metal has been removed by a cutting process.

leading edge. The foremost edge of an airfoil or propeller blade.

lightening holes. Holes in aircraft parts where metal has been removed because the metal in that specific location was not doing any useful work and its removal lessened the weight with no decrease in the structural strength of the part; such holes are usually flanged or stiffened at their edges.

longeron. A longitudinal member of the framework of an airplane fuselage or nacelle.

loom. The web sections of wing ribs.

lucite. Trade name for a transparent plastic material. Used for windows, nose, etc., on airplanes, and for tooling purposes. Softens when heated.

magnetic inspection. The determination of flaws at or near the surface of a ferro-magnetic metal by magnetizing the metal or member to be tested and then treating the surface with fine, iron particles.

main beam. A rigid body designed to transmit transverse loads in shear and/or bending to its points of support.

manifold. A pipe with outlets or branches. In welding, it generally refers to a pipe to which several cylinders of gas may be connected to supply gas to a number of blowpipes. A pipe having several branches which leads the fuel-air mixture to the cylinders of an engine or leads the exhaust gases from the engine.

masonite. Trade name for an alkaline plastic made from wood pulp.

melting point. The temperature at which a substance changes from the solid to the liquid state.

member. Any essential part of a structure or machine.

metal. An element which is typically fusible and opaque, is a good conductor of electricity, and has a typical metallic luster.

mold. The cavity in which anything is shaped. A frame or body on or about which something is made.

mold line. A line formed by the intersection of two planes (flat surfaces). In the case of an angle where the bend is a radius, the mold line will be in space at the point where the outside surfaces or the legs of the angle would meet if they were extended.

monel metal. An alloy of 67 per cent nickel, 28 per cent copper, and 5 per cent iron, manganese, silicon, and other metals. Monel metal is extremely resistant to corrosion.

monocoque fuselage. A type of fuselage construction wherein the structure consists of a thin shell of wood, metal, or other material, supported by ribs, frames, belt frames, or bulkheads, but usually without longitudinal members other than the shell itself. The structure is built so that the shell carries the greater part of the stresses to which the fuselage is subjected.

motor mount. A support on which to mount the motor of an airplane.

nacelle. An enclosed shelter for passengers or for an engine. A nacelle is usually shorter than a fuselage and does not carry the tail unit.

neutral flame. The neutral flame is created by burning acetylene with oxygen in such proportions that all particles of carbon and hydrogen in the fuel gas are consumed. It produces a temperature of approximately 6300° F.

nose. The bow of an airplane, usually the front of the fuselage or hull. It also refers to the front end of a wing rib or the leading edge of a wing.

oil can. (Colloquial.) An expression meaning that the skin is slightly bulged up between rows of rivets on a metal-skin structure. This bulge can often be pushed back and forth by applying a slight pressure to the metal so that the so-called "oil can" will react like the bottom of an oil can. Such a condition is due to the metal being forced out of line by the row of rivets and must be avoided.

oxide. Coating or scale formed on metal when combined with oxygen.

oxidizing flame. A flame containing an excess of oxygen which combines with the metal. This action is damaging to many metals.

oxy-acetylene welding. A gas welding process wherein the welding heat is obtained from the combustion of oxygen and acetylene.

oxygen. An odorless, colorless, tasteless gas, forming approximately 21 per cent of the atmosphere.

peening. Mechanical working of metal by means of hammer blows.

penetration. Depth of fusion obtained in a welded joint.

pickling. The removal of surface contamination from metals or alloys by dipping in a solution.

plastic. A general term applied to low-density, synthetic resins and materials in solid form which can be formed or molded, under suitable conditions, to desired shapes. The chemical ingredients of plastics unite by means of polymerization.

plexiglas. Trade name for a transparent plastic material. Used for windows, nose, etc., on airplanes, and for tooling purposes. Softens when heated.

porosity. The presence of gas pockets or inclusions.

power brake. A sheet-metal brake operated by power.

preheating. Heat applied prior to welding, cutting, or forming operations.

pressure gauges. The instruments used to indicate the pressure of the gases as they flow toward the blowpipe.

primer. Paint used for a first, or prime, coating. Zinc chromate is usually used as a primer in airplane construction.

production illustration. A drawing in perspective to illustrate a part or an assembly.

ream. To widen the opening of a hole; to bevel out. To enlarge a hole with a reamer.

rib. A light, curved, metal part mounted in a fore-and-aft direction within a surface. The ordinary ribs give the surface its camber, carry the fabric, and transfer the lift from the fabric or skin to the spars.

rig (airplane). To assemble, adjust, and align the parts of an airplane.

ring cowling. A ring-shaped cowling placed around a radial air-cooled engine to reduce its drag and improve cooling.

rivet draw. A special bar with a hole in one end large enough to slip over the rivet shank which is used to draw or set the sheets together.

rivet gauge. A tool for inspecting the height and diameter of a rivet shank after upsetting or driving. A flush-type rivet gauge is used for inspecting the height of the manufactured head of the rivet projecting above or below the surface.

rivet set. A tool to drive or upset rivets. It is a tool with a cup-shaped depression in one end that must fit the head of the rivet being used.

riveting. The method of permanently fastening or joining sheets and parts together with rivets.

riveting, blind. Upsetting rivets from one side only.

riveting, gang. Upsetting two or more rivets at a time with a machine riveter.

riveting, gun. A compressed-air or pneumatic hammer used for riveting.

riveting, hand. Upsetting rivets by hand hammer or hand squeezer.

setting. The bucker holding the bucking bar alongside the projected rivet shank to draw the sheets together before upsetting the rivet.

shear beam. An auxiliary beam designed to take local shear loads and transmit them to portions of the structure where they can be properly absorbed.

shearing. Cutting metal by means of shears of various types.

silver soldering. A brazing process wherein a silver alloy is used as a filler metal.

skin. The outer covering of an airplane.

solder. A metal or metallic alloy used, when melted, to join metallic surfaces. Solders which melt readily are soft solders; others fusing at a red heat are hard solders.

span. The maximum distance, measured parallel to the lateral axis, from tip to tip of an airfoil, of an airplane wing inclusive of aileron, or of a stabilizer inclusive of elevator.

specific gravity. The ratio of the weight of a substance to the weight of an equal volume of water at some standard temperature.

spinner. The fairing of approximately conical shape which is fitted over the propeller hub and revolves with the propeller.

splice. The joining of the ends of two pieces of rope, wire, wood, or metal in such a manner that the splice (joint) is just as strong as the material spliced.

spot weld. A weld, either gas or electric, which is not a continuous seam weld but is made only at spots. Spot welding generally refers to electrical resistance welding.

stainless steel. The term, stainless steel, refers to a series of alloys containing chromium and nickel. These alloys have special properties such as high resistance to corrosion and strength at high temperatures.

strain. The effect of the application of external forces to a body.

strength. The power to resist force.

stress. The resistance to distortion within a body. It can only exist in unconfined bodies when these are solid or plastic.

strut. A compression member of a truss frame, for instance, the vertical members of the wing truss of a biplane (interplane struts) and the shorter, vertical and horizontal members separating the longerons in the fuselage.

subassembly. An assembly of component parts to make up a section of an airplane.

swage. A machine or device for drawing metal.

tailboom. A spar connecting the tail surfaces and the main supporting surfaces of the airplane.

tap. A tool for forming an internal screw thread. In an electric circuit, a point where a connection may be made.

template. A full-sized pattern from which structural sections are marked out. It may be paper, cardboard, plywood, or metal.

tensile strength. The resistance to forces tending to pull the material apart.

thermit welding. A nonpressure welding process wherein the heat is obtained from liquid steel produced by a thermit reaction and the filler metal is supplied by the steel produced in this reaction.

trailing edge. Rearmost edge of an airfoil or of a propeller blade.

upsetting. The operation of setting or driving the rivet shank to the correct height and diameter.

vibrator. A compressed-air or pneumatic hammer used for riveting.

warp. Any variation from a true or plane surface.

web. A structure member used to give form or to transmit load.

wing. The main lifting and supporting surface of the airplane in flight, designated as right and left.

wing rib. Chordwise member of wing structure of airplane used to give form to the wing section and to transmit the load from the skin to the spars.

wing tip. Outer or outboard end of a wing.

workability. The degree of ease and smoothness of cut obtainable with hand or machine tools.

yield strength. The point at which permanent deformation in the structure begins to take place.

VIII HEAT TREATMENT OF SHEET METAL

The physical properties of most metals and alloys can be changed by heat treatment. Heat treatment consists of heating and cooling under various conditions.

Heat treatment not only affects the grain structure of the metal, but may affect its chemical composition. As a rule, the longer the cooling time of a metal or alloy which has been heated above its critical temperature, the larger the grain structure. A few of the metals become softer when cooled rapidly, but most become harder. The critical point is the temperature to which the metal must be heated to allow change in the crystalline structure. At ordinary temperatures, changes in crystalline structure may take place, but the process is slow. Most metals require heating to some degree of redness to reach the critical point. At this temperature, crystalline structures may change rapidly. Aluminum shows no changes in color when heated.

With proper heat treatment, such characteristics as hardness, toughness, springiness, ductility, and workability may be controlled. Usually the heating is brought about gradually, and the metal is held at the required temperature for varying lengths of time.

When a substance must be cooled rapidly, it is quenched. Quenching is submerging the heated metal into some liquid such as salt solution, plain water, or oil. These liquids may be at any predetermined temperature. A cold spray or air blasts are sometimes used.

Annealing, hardening, tempering, and normalizing are forms of heat treatment.

Special characteristics are often brought out by heating to above the critical temperature, followed by cooling to a predetermined temperature, and then sudden cooling by quenching. Quenching is usually done on a rising temperature. The type of quenching or heat treatment used is largely determined by the properties desired in the finished product.

Special heat treatments such as carburizing, cyaniding, case hardening, and nitriding are sometimes used. Heat treatment can be used to relieve stresses and strains within the metal, which have been set up by hammering, rolling, working, welding, and other operations to which the material may have been subjected.

Annealing. Annealing is a form of heat treatment which usually consists of heating the metal slightly above the critical temperature for a predetermined length of time and then cooling it slowly. Cooling is

Fig. 52. Quenching extruded aluminum alloy parts after heat treatment. (Courtesy Aluminum Company of America)

sometimes carried out in a furnace by a gradual lowering of the temperature. However, cooling may be carried out by burying the part in slaked lime, sand, mineral wool, glass wool, shredded asbestos, or ashes. Steel and other alloys of iron usually become softer when annealed. Small iron and steel parts are often annealed by heating to a red temperature with a blowpipe. After the proper temperature has been reached, the part is buried in sand or lime and allowed to cool slowly. This treatment relieves internal strains and softens the metal. It is usually necessary to heat-treat parts which have been formed, as the working of the metal sets up strains and may harden it.

Hardening. Most parts are hardened after forming by special heat treatment. This hardening consists of heating the part slowly to a point above the critical temperature. This temperature is maintained until the parts are thoroughly heated and the crystalline structure has had time to stabilize itself. The part is then quenched in brine, water, or oil. Oil is often used because the quenching action is not as severe as that of water or brine. Steel tools, such as cold chisels, when hardened are said to have been tempered. Heavy parts having large cross sections are hardened to a greater extent on the outside surfaces than on the inside due to the slower cooling beneath the surface. The parts being

Fig. 53. Quenching extruded parts after heat treatment. (Courtesy Reynolds Metals Company)

quenched should be moved around in the quenching solution. This helps relieve stresses and strains on the surface of the part. If a part is allowed to remain quietly in a quenching solution, small cracks or checks may appear on the surface.

Normalizing. Normalizing is similar to annealing. The material being normalized is heated uniformly throughout at a temperature slightly above the critical point. It is then allowed to cool in still air. Normalizing will produce a hard, strong metal which is less ductile than when annealed. Normalizing relieves internal stresses and produces greater uniformity of structure throughout the piece being treated. Aircraft steels are usually normalized. Annealed parts are not commonly used in aircraft structures because of their softness and lack of strength. Normalizing reduces the chance of failure due to fatigue. Chromium alloys are normalized before the regular hardening operations and heat treatments are carried out.

Case Hardening. Case hardening is a process by which the outer layer of a steel part is hardened. This is done where the surface will be subjected to severe wear. Case-hardened material has an extremely hard outside which tends to be brittle. This outside is supported by a softer, tougher core which gives strength to the part. The part being case hardened is submerged in either a liquid or finely divided solid, from which extra carbon is absorbed. While in contact with the liquid or finely divided solid, the part is heated to the necessary temperature. A molten bath of cyanide, scraps of leather, or other substances containing carbon may be used. After case hardening, the part may be heat-treated to bring about the proper condition of hardness and temper.

Carburizing. Carburizing is a form of case hardening and consists of packing the part in a steel box surrounding it by a mixture of some such substance as ground bone, charred leather, coke, wood, charcoal, or even sugar. These substances may be used alone or in combination. Barium and sodium carbonate are often added to assist in the reaction. Only small amounts should be added. A typical carburizing mixture contains approximately 60 per cent charcoal, 25 per cent powdered coke, 14 per cent barium carbonate, and 1 per cent sodium carbonate. The parts are packed in an alloy steel box made of corrosion-resistant material. The packing material is arranged so as to be in close contact with the part to be hardened. Copper plating will protect any portion of the part from the hardening process. The box is sealed with moist fire clay. If the fire clay tends to crack, a little salt may be added. The whole is then heated in a furnace to a temperature between 1600° and 1700° F. The temperature is maintained at this level until the desired results are obtained. The longer the parts are held at this temperature, the deeper the carburizing will penetrate. A case-hardened layer $\frac{1}{64}$ in. in depth is usually obtained in 1 hour of heating at the above temperatures. Another method of carburizing makes use of a melted salt to which amorphous carbon has been added. Carbon must be added at intervals to replace that driven off by the heat. A depth of from $\frac{1}{10}$ to $\frac{2}{10}$ in. of hardening may be obtained per hour by this method. Hardening in a furnace may be carried out by passing methane gas through the furnace while the part is held at the desired temperature. Small areas may be hardened by the application of an oxy-acetylene flame having a large excess of acetylene.

Cyaniding. Cyaniding is a form of case hardening in which the material to be hardened is heated in contact with a cyanide salt. All cyanide salts are extremely poisonous, and great care must be used to avoid coming in contact with the salt or breathing the fumes. The parts to be hardened are heated and dipped into the melted salts.

Nitriding. Nitriding is a form of surface hardening used for special alloys containing aluminum in combination with chromium or molybdenum. The parts to be hardened are heated in a special furnace to temperatures ranging from 900° to 1000° F. Ammonia is introduced into the furnace and, at this temperature, it breaks down and releases an active form of nitrogen which combines with the outside layer of the part being treated. This outside layer is extremely hard and brittle. The nitrided layer is harder than the surface obtained from treating with carbon. The nitriding process does not cause cracking or distortion of the part. No quenching is required, and usually no scaling occurs. The nitrided area is resistant to corrosion. Any part of the surface may be protected from the nitriding operation by coating it with tin or solder.

The extensive use of aluminum in the industries was brought about by the discovery of methods of heat treating its alloys. Many of the alloying substances used in aluminum alloys dissolve much more freely in the aluminum at high temperatures. In the heat treatment of aluminum alloys, the alloy is heated to a temperature just below the melting point so that as much as possible of the alloying substances may go into solution with the aluminum. This dissolving process takes place in what is called a solid solution. After staying at the proper temperature for the desired length of time, the alloy is quenched in water. This causes the alloy to become solid, holding in solution the dissolved alloying substances. In this condition the alloying substances are so finely divided that their particles are not visible even under a high-powered microscope. This is known as "solution heat treatment."

After quenching, an aging time is usually necessary in order that the alloy may take on its permanent characteristics. If this aging is carried out at a temperature above the normal air temperature it is called "precipitation heat treatment." Under these conditions, part of the dissolved alloys separates out in the form of extremely fine particles of the alloying substance. The particles remain embedded in the aluminum and seem to prevent slippage along the crystal planes of the metal. This is something like putting sand on a slippery sidewalk to prevent slipping. This separating out of fine particles of the alloying substance may take place at room temperature. This is called the "natural aging process." The alloys of aluminum designated as 17S and 24S usually

Fig. 54. Distortion of aluminum alloy sheet caused by heat treatment. (Courtesy Reynolds Metals Company)

age at room temperature. The aluminum alloys designated as 51S and 53S are aged at a temperature of about 300° F. The aging period may range from a few hours to several days.

Aluminum alloys which have been cold-worked become strain hardened. This condition may be removed by annealing. Pure aluminum, such as is indicated by 2S, becomes hard when worked cold. This hardness disappears almost at once when the metal is heated to a temperature which permits recrystallization. For 2S aluminum this temperature is approximately 650° F. For 3S and 4S aluminum, a temperature of around 750° F. is necessary. Heating for longer periods of time at lower temperatures will accomplish the annealing. The rate of cooling of aluminum after annealing has but little effect on its properties. The cooling should not be too rapid.

In the heat treatment of the aluminum alloys, great care is required to obtain the desired results. The alloy must be raised to a temperature which will permit recrystallization in order that the strain-hardening condition be removed, but the temperature must not be high enough to destroy the properties derived from the regular heat treatment. In other words, the strains set up by the cold working may be removed independently of the hardness and strength of the material, due to the heat treatment. Heating to too high a temperature would completely soften the material and require reheat treatment. The heating of these alloys to a temperature between 630° and 650° F. will remove the hardness brought about by cold working.

When it is desired to anneal completely the alloy, it should be heated to a temperature of approximately 775° F. for about 2 hours. The alloy is then allowed to cool slowly in the furnace to a temperature of approximately 500° F. The rate of cooling should not exceed about 50° per hr. After the material has been cooled to approximately 500° F., it may be removed from the furnace and allowed to air-cool to room temperature.

Most aluminum sheet used to form aircraft parts is in the annealed condition. After the part is formed, either by machine or by hand, it is necessary to heat-treat it in order to bring about the desired condition of strength and hardness. Parts of this kind are usually heated for the required length of time in a chemical bath to the required temperature. As soon as the part is removed from this bath, it is quenched in cold, running water. The heat treatment of formed parts usually causes some deformation of the part, requiring handwork to restore it to its proper shape.

The various heat treatments, such as case hardening, carburizing, and cyaniding, in which the addition of carbon to the metal affects the hardness are not used with aluminum and aluminum alloys. The properties of aluminum brought about by heat treatment are due almost entirely to grain structure. This heat treatment may be carried out either in a salt bath or an air furnace. However, the temperature of the salt bath may be more easily controlled. The formed parts are submerged in the salt bath, which consists of molten sodium nitrate, and immediately quenched in cold, running water after removing from the bath. All salt from the heat treating solution must be washed from the metal after quenching to prevent corrosion. Care should be taken that damp or wet articles or water do not come in contact with the hot salt solution as an explosion may result.

The time which the material remains in the salt bath depends upon its thickness and the composition of the alloy. For material 0.020 in. or less in thickness, a 10- to 15-minute period in the salt bath is sufficient. For material 0.25 to 0.50 in. in thickness, a period of 45 to 60 min. is required. After solution heat treatment the parts should be immediately quenched in cold, running water from a temperature which should be no lower than approximately 950° F.

The temperature of the quenching water should be as low as practicable and should not exceed 100° F. at any time during the quenching period. Sudden cooling during quenching tends to warp the material. Most parts must have additional forming after being heat treated.

The various alloys require slightly different methods of heat treatment. For example, 24S reaches its maximum strength in the "T" condition within four days after solution heat treatment and quenching, when maintained at room temperature. The alloy, 75S, reaches only about ⅔ of its required strength after one month's aging at room temperature. This alloy requires artificial aging to bring about its maximum strength. The highest properties result from aging at 250° F. for 24 hrs., starting within 2 hours (or *after* 2 days) after quenching. Most aluminum alloys do not acquire their maximum hardness immediately after quenching, but require an aging period. This aging period may be extended greatly by refrigerating the parts immediately after quenching. Rivets or other parts may be kept for days or months in their softened condition, depending upon the temperature of the refrigerator.

The annealing and heat treatment of alloys such as stainless steel are quite different from those processes used for aluminum alloys. Annealing and heat-treatment operations accomplish one or more of the following results: (1) restore corrosion-resisting qualities which may have been destroyed by working at high temperatures or by welding; (2) relieve stresses set up in the material by cold-working; (3) obtain special physical properties in alloys which respond readily to heat treatment.

Certain chrome nickel alloys are not stable when heated to temperatures between 800° and 1500° F. When these alloys have been heated to temperatures within this range, they must be re-annealed at a temperature of approximately 2000° F. to restore their original properties. The material must then be cooled rapidly to a temperature below 800° F. It is necessary to reheat-treat these alloys after cold-working or forming to relieve the stresses set up in the material. Most of these alloys may be softened for forming by heating to approximately 1450° F. for a short time and cooling in still air. The temperature at which these alloys are annealed or heat-treated varies with the alloy and should be checked carefully with the manufacturer's specifications.

Fig. 55. Anodizing aluminum alloy parts. (Courtesy Douglas Aircraft Company, Inc.)

After forming, many of the aluminum alloys are anodized before being placed in aircraft structures. This is particularly true if the material is to be painted or have other finishes applied. Anodizing consists of forming an artificial oxidation on the surface of aluminum or magnesium alloys. The material to be anodized is submerged in a chemical solution and treated with an electric current. The surface thus formed resists corrosion and forms a good bond for paints or other finishes. Either a sulphuric acid or chromic acid electrolyte is used.

IX TOOLS AND MACHINERY

Sheet metal is formed by passing the cast ingots through heavy steel rolls which press the ingots into sheets of the desired thickness. The sheets must be sheared to the proper size and bent, twisted, hammered, stretched, shrunk, pounded, sawed, or treated in some way before they may be fitted into the airplane structure. It is the responsibility of the sheet-metal worker to take the raw material and form the finished part.

The hand tools of the sheet-metal worker are simple and not many in number. The sheet-metal worker's kit should include layout tools such as shown in Figure 56. Figures 57, 58, 59, and 60 illustrate some of the hand tools used in forming sheet metal.

Most hand forming is done by the use of pounding tools. Steel hammers vary from those having flat polished faces to those that are high crowned or rounded on the face. Mallets of many different shapes and materials, such as hard and soft woods, plastics, Dural, rawhide, lead, and rubber, are used in sheet-metal forming. The pounding tools of this type are used to form hollows, flanges, corrugations, and beads, as well as to stretch and shrink the metal.

It is necessary that the sheet-metal worker know not only the force with which the blow is to be struck, but also the direction and the exact point on the sheet which is to receive the force of the blow. The material may be hardened by hammering, and the grains of the metal may be made to move over each other, piling up and making the material thicker, or spreading out and thereby thinning the material.

Bumping is used to form hollows and irregular curves in a sheet. Bumping may be done over a form block by the use of a sandbag and various kinds of mallets and with male and female dies. Much of the hand bumping which was done in the early days in aircraft manufacturing is now performed by drop hammers and other sheet-metal-forming machines. Bumping requires a high degree of skill. It is often necessary for the airplane mechanic to make use of this process.

Bending, particularly where the bend is along a straight line, is usually done with the aid of a bending brake. Much bending, however, is done by hand, such as the forming of flanges or curved parts. In bending,

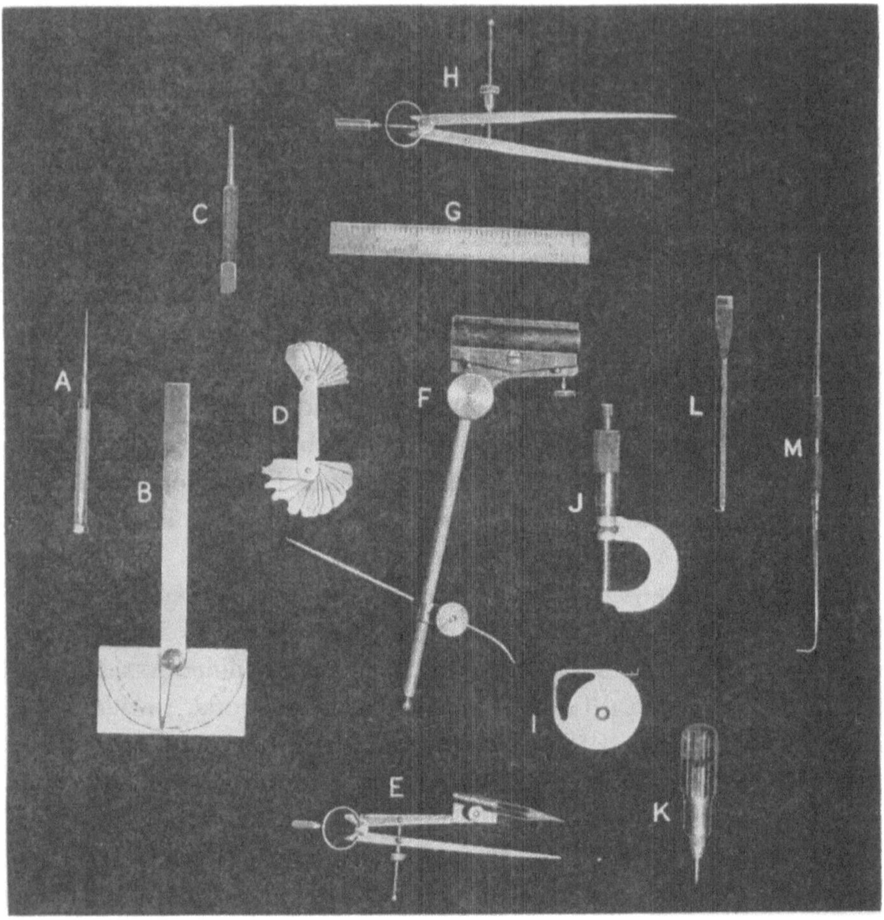

Fig. 56. Layout tools: (A) taper punch; (B) protractor; (C) center punch; (D) feeler gauge; (E) pencil compass; (F) trammel gauge; (G) steel rule; (H) dividers; (I) steel tape; (J) micrometer; (K) small taper punch; (L) cold chisel; (M) scriber. (Courtesy The Glenn L. Martin Company)

allowance must be made for the thickness of the metal and the sharpness of the bend which the metal will stand. Some metals are ductile and will stand bending back on themselves without fracturing. Others require a bending radius equal to several times the thickness of the metal. It is necessary that the aircraft sheet-metal worker be thoroughly familiar with the bending properties of the metal with which he is working and at all times make proper bend allowances.

Spinning, while not used as extensively as formerly in the aircraft industry, consists of forming a sheet of metal while it is revolving at a comparatively high speed in a lathe or other device. The metal may be

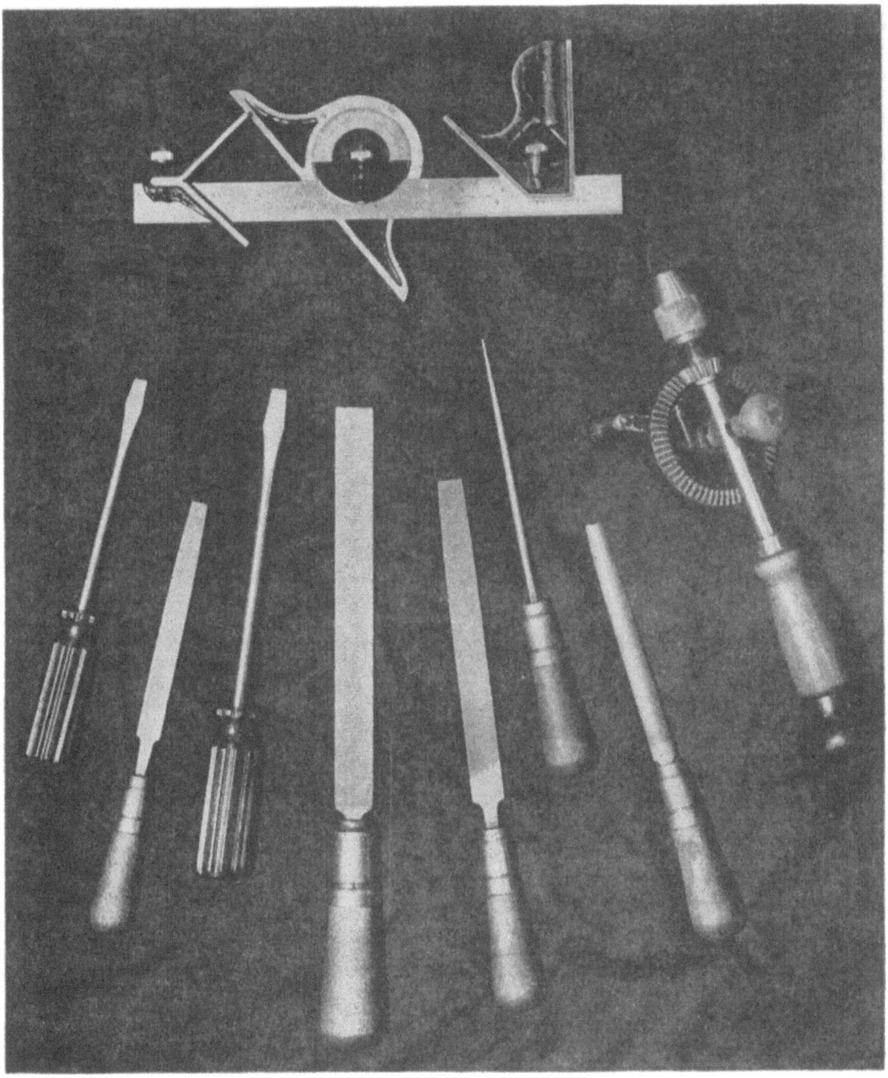

Fig. 57. Hand tools: combination square; screw drivers; assorted files; taper punch and hand drill. (Courtesy The Glenn L. Martin Company)

formed over a rotating form block or be "free spun," much as a potter forms his clay on the revolving potters' wheel. The spinning tools may consist of a stick of wood having a rounded end which is forced against the revolving metal, or steel rods having a polished, rounded, spoon-shaped end. While the tools themselves are simple, the art of spinning requires a high degree of skill.

There are several different methods by which sheet metal is cut to the proper shape. The hand snips, which are simply heavy, short-bladed

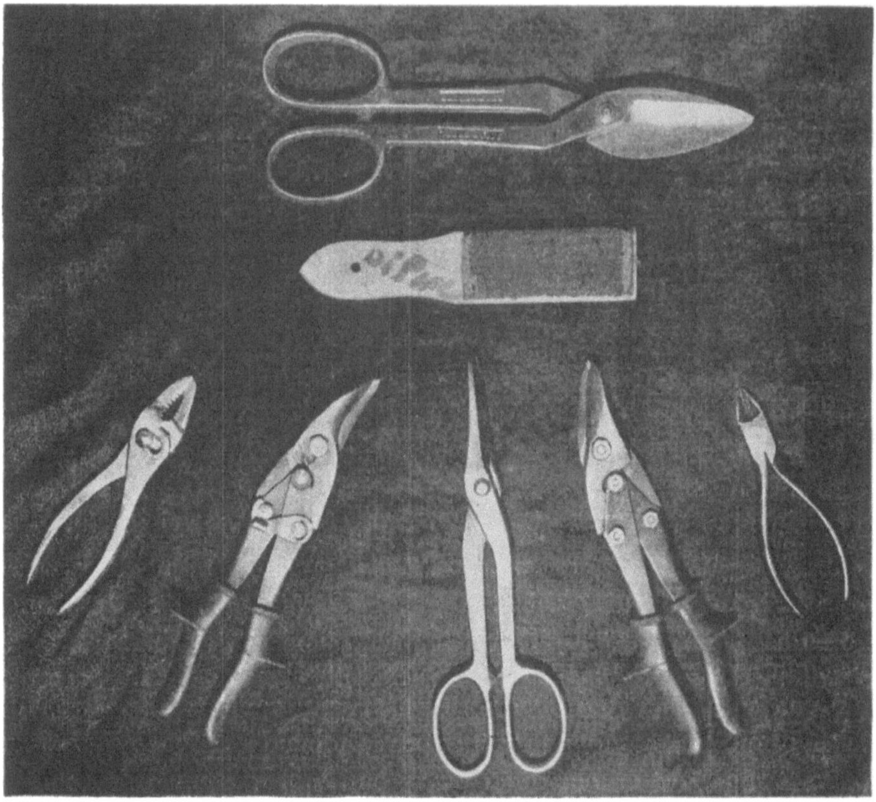

Fig. 58. Hand tools: straight metal snips; card for cleaning files; adjustable pliers; assorted airplane snips. (Courtesy The Glenn L. Martin Company)

shears, are a requirement of all handworkers in sheet metal. The snips may be either lefthanded or righthanded, according to the edge of the sheet on which they are to be used. Aviation, sheet-metal snips may have curved blades for cutting along curved lines. Power shears range from a few inches to several feet in length and are commonly used in cutting sheet to the desired dimensions. Most sheet metal may be sawed by hacksaws, circular saws, or band saws. Odd shapes may be cut with the proper snips, routers, or nibblers, and power cutters of other types.

Much material is cut to shape by power presses equipped with a punch and die. Holes may be punched or drilled. It is impossible to draw a distinct line between hand forming and machine forming, as many hand-forming operations are carried out with the aid of machines. Perhaps the one essential difference between hand forming and machine forming is the fact that hand forming always requires judgment. The machine can only perform operations for which it is set. The handworker

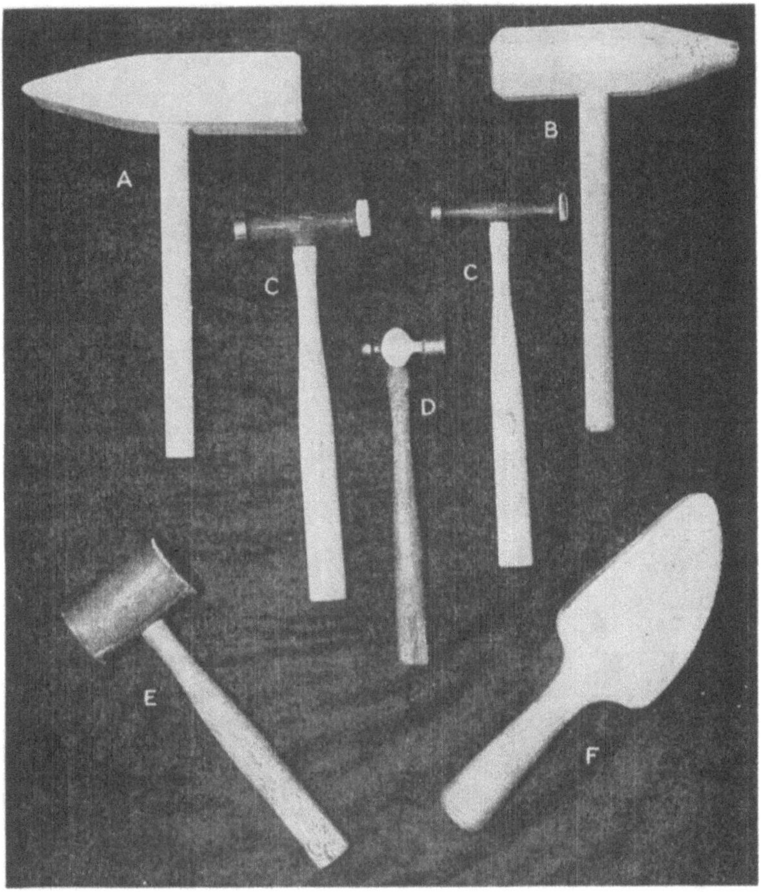

Fig. 59. Hand tools: (A) cross peen wood mallet; (B) round peen wood mallet; (C) planishing or finishing hammers; (D) ball peen hammer; (E) rawhide mallet; (F) wood spanker. (Courtesy The Glenn L. Martin Company)

can make small corrections and perform operations which require varying degrees of adjustment. For example, a part formed by a machine may lack just a little forming to make an exact fit to the check block or form. The handworker can take this part, open an angle a little, bend the part slightly, twist it a little, and obtain an exact fit.

Machinery is used extensively by the aircraft manufacturers when large numbers of identical pieces are to be made. Machinery has displaced the preliminary hand forming of almost all parts of the aircraft

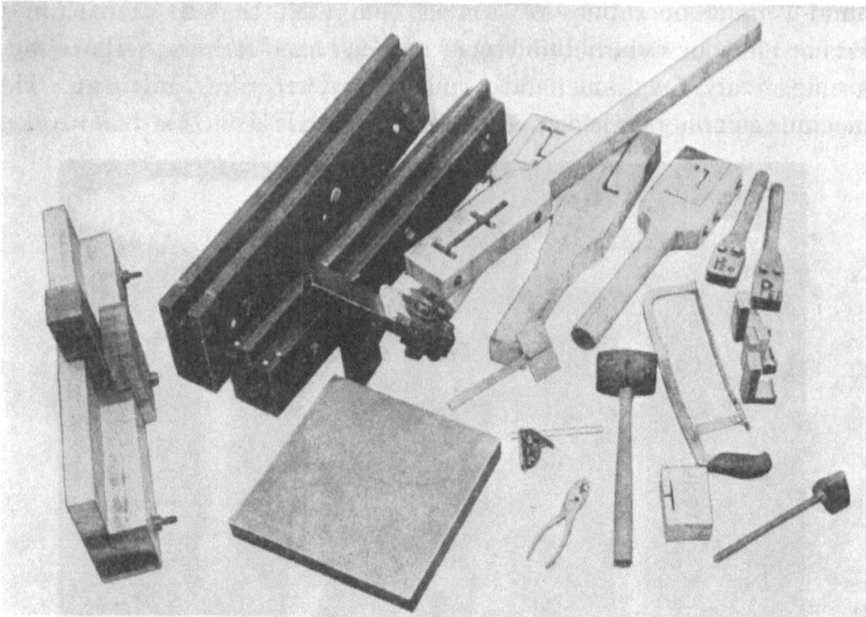

Fig. 60 (a).

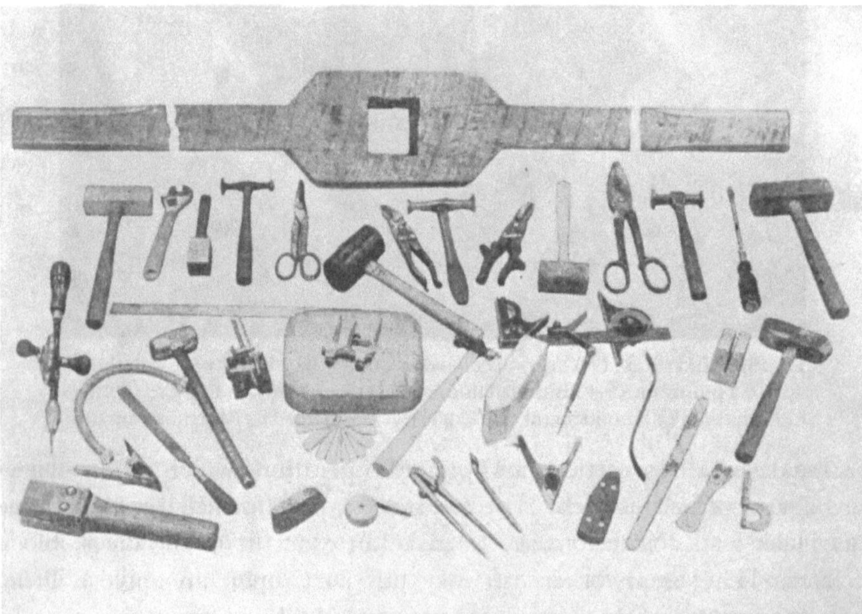

Fig. 60 (b).

Figs. 60 (a), 60 (b), and 60 (c). Hand-forming tools. (Courtesy The Glenn L. Martin Company)

structure. Sheet metal is rolled both to flatten it and to form it into curves of various degrees. Not only may cylindrical curves be formed, but tapering curves, beads, and corrugations may be made by means of rolls.

Parts made of sheet may be stretched or shrunk by means of special machines. In stretching, the two edges of the sheet are seized between the jaws of the machine, and the part is stretched the desired amount.

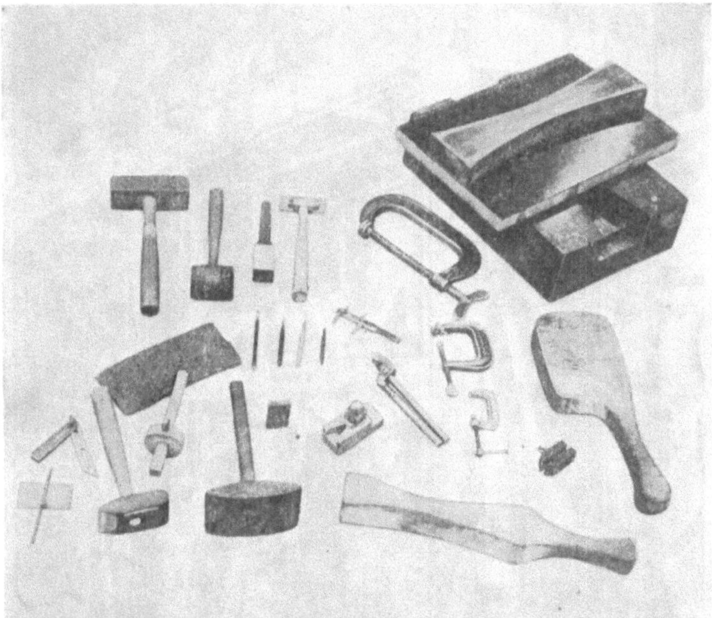

Fig. 60 (c).

Sheet metal is made around various forms by being pulled into shape by a machine. Sheets used for engine-cowling and fuselage curves may be formed by stretching over a properly shaped die. In stretching, the metal is pulled beyond its elastic limit and takes a permanent set. When this process is used for aluminum alloys, it does not seem to weaken the material.

Drop hammers are particularly valuable in forming curves. The drop hammer is a rapid means of producing large numbers of identical parts.

Hydraulic presses form parts from sheet material under enormous pressures, using rubber pads to distribute the pressure and force the sheet to conform to the die. The Yoder hammer, which is a pneumatic hammer, takes the place of hand pounding in forming curves in sheets.

Corrugating is done by means of large presses, one corrugation being formed on each stroke of the press. Huge shears are used to cut large sheets into the desired shape. Large presses are used to blank out sheet, making parts of various sizes and shapes.

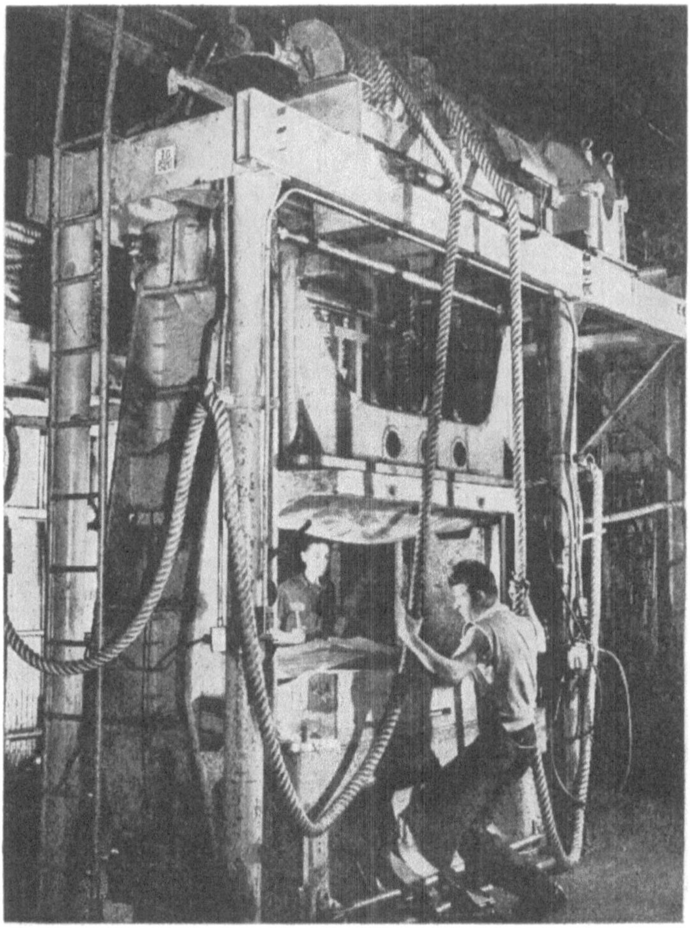

Fig. 61. Operating a rope-controlled drop hammer. (Courtesy Douglas Aircraft Company, Inc.)

When using any machine to form sheet metal, it is necessary for the machine operator to have a good understanding of the processes involved and the characteristics of the metal which is being formed. In many of the large factories it is necessary for the operator of each machine to set up his own punch and dies and to make all adjustments necessary to form the part correctly. For example, the drop hammer is not entirely automatic. It is necessary for the operator to use a high degree of skill

Fig. 62. Airplane parts being formed on a bending press. (Courtesy The Glenn L. Martin Company)

Fig. 63. Shearing sheets to size in the stockroom of a large aircraft manufacturer. (Courtesy The Glenn L. Martin Company)

Fig. 64. A router being used on an extruded part. (Courtesy Douglas Aircraft Company, Inc.)

Fig. 65. A radial arm router. (Courtesy The Glenn L. Martin Company)

Fig. 66. A radial arm router and drill forming parts by the use of a template. (Courtesy Douglas Aircraft Company, Inc.)

Fig. 67. A band saw used to cut metal. (Courtesy The Glenn L. Martin Company)

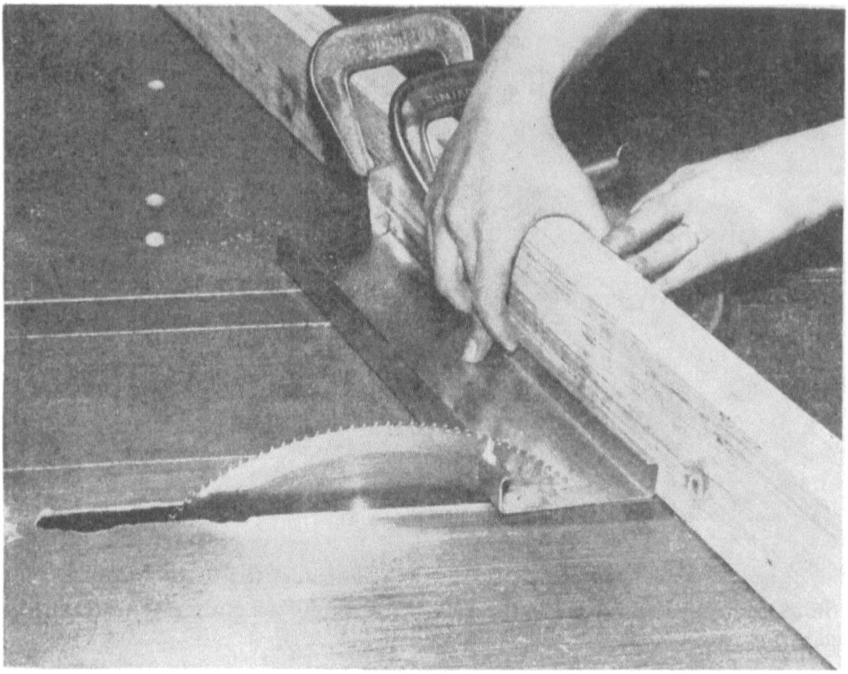

Fig. 68. Cutting metal with a circular saw. (Courtesy The Glenn L. Martin Company)

Fig. 69. Forming sheet metal on a hand-operated slip roll. (Courtesy Douglas Aircraft Company, Inc.)

Fig. 70. Forming sheet metal in a large power-operated slip roll. (Courtesy Douglas Aircraft Company, Inc.)

Fig. 71. A gang piercing machine to make a number of holes in one operation. (Courtesy The Glenn L. Martin Company)

Fig. 72. A stretching machine used to form airplane parts. (Courtesy The Glenn L. Martin Company)

Fig. 73. A large hydraulic press using the rubber-pad method for blanking and forming. Note that there are four tables on which work is prepared, thus permitting rapid operation of the press. (Courtesy Aluminum Company of America)

Fig. 74. A modern drop hammer. (Courtesy Douglas Aircraft Company, Inc.)

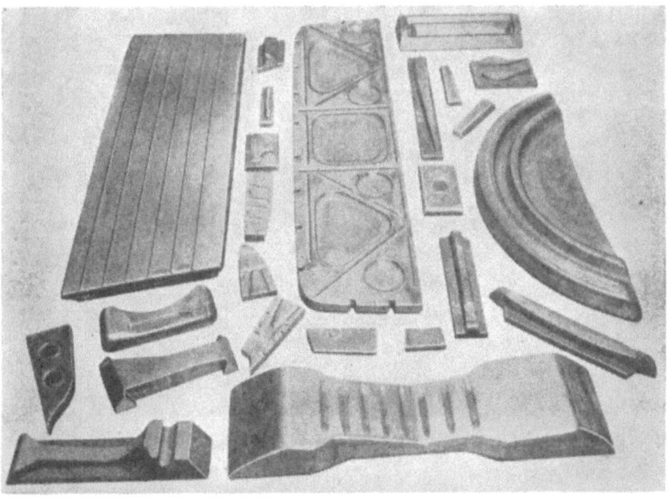

Fig. 75. Tools for forming aluminum by the rubber-pad method. These tools are placed on the press table, under the sheets to be formed. (Courtesy Aluminum Company of America)

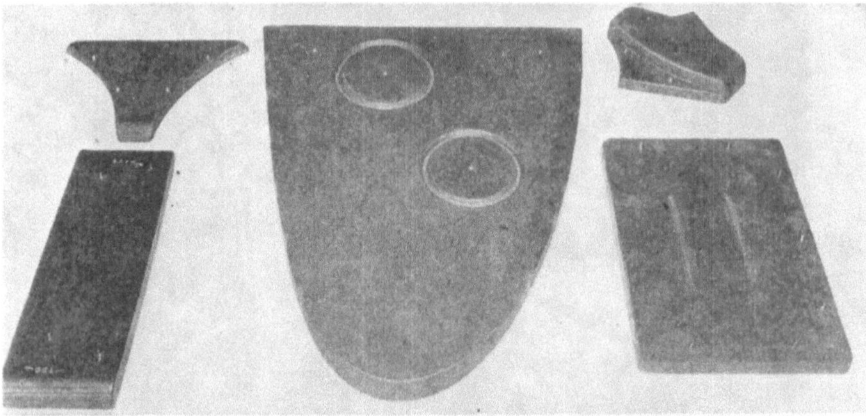

Fig. 76. Wood fiber tools for forming aluminum parts on the hydropress. (Courtesy Aluminum Company of America)

Fig. 77. A yoder hammer. (Courtesy The Glenn L. Martin Company)

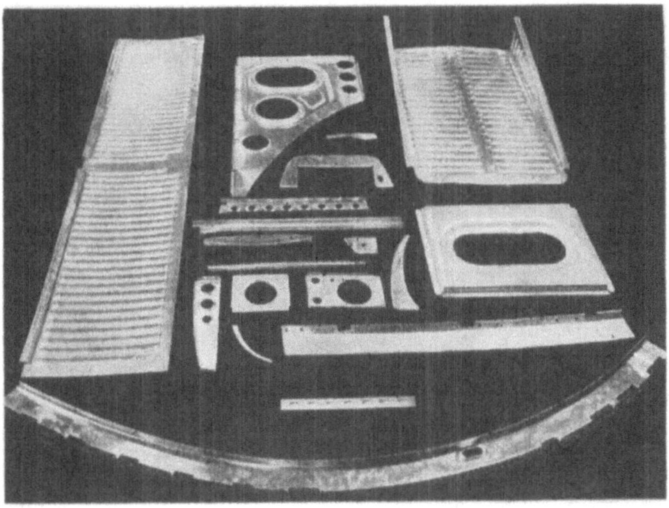

Fig. 78. Airplane parts formed on a hydropress by the rubber-pad method. (Courtesy Douglas Aircraft Company, Inc.)

Fig. 79. Forming aluminum with a punch and die. (Courtesy Aluminum Company of America)

and judgment in deciding how hard a blow to strike. In the formation of many parts it is necessary to use rubber pads to distribute the pressure applied to the sheet.

Form blocks, backing blocks, and check blocks are of great importance to the handworker in sheet metal. A form block is made of a material which will resist pounding, such as hardwood, plastic, compressed wood, or metal. The sheet metal is shaped over the form block. Allowance must be made in the preparation of the form block for the metal's springing back. A backing block is often used to clamp the sheet of metal to the form block. Check blocks are used to check the fit of the finished part. The form block should not be used to make the final check, as form blocks may become distorted and deformed. The sheet-metal worker should be able to prepare his own form blocks and backing blocks. The check block is usually obtained from the tool or supply room. A check block is not usually made if only a few parts are to be formed, such as might be done in making a repair. The old part may be used to check the newly formed part or the new part itself may be fitted into the structure.

X GENERAL SAFETY PRACTICES

The sheet-metal worker should be thoroughly familiar with all safety practices in connection with his work. There is always a chance of accidents occurring in any trade through carelessness or ignorance.

The following are the general safety practices which should be observed by the worker in sheet metal.

Consider safety as a part of every job you do.

Take time to do your job the safe way.

Don't engage in any kind of horseplay on company property.

When crossing aisles and passageways, walk with care, keeping to the right, especially at intersections. When using stairs, keep hands out of pockets, ready to catch yourself if you should stumble.

Place no tool, material, or other object in any place where it might fall and injure someone or fall into machinery.

Don't toss tools or parts to fellow workmen.

Never walk under a ladder. There is always the possibility of being struck by something from above.

Don't use unsafe ladders or improvised ladders, such as packing boxes, etc.

Don't transfer from one ladder to another. Get down.

Danger tags are placed on machinery and other equipment for a definite purpose and are not to be removed.

Handle waste, rags, and industrial towels with care because of the possibility of concealed sharp objects.

Oily waste, rags, and gloves should always be placed in containers provided for that particular purpose. Otherwise, they may ignite and cause a fire.

Clean up spilled oil at once, no matter who spilled it.

Unless your job involves repairing and maintaining electrical machinery, wires, and apparatus, do none of this kind of work.

To avoid strains, learn to lift the right way: bend your knees, keep your body erect, then push upward with your legs. It is much easier and safer. Do not try to lift or push objects which may be too heavy for you. Ask for help when you need it.

Never wear loose clothing, dangling neckties, or loosely woven clothing such as sweaters when working near rapidly revolving machinery.

In the factory, women must wear low-heeled shoes with closed heels and toes. A durable, lightweight cap with a visor is worn by women working on machines, welding, riveting, or drilling, or doing any other work where a hazard to their hair or scalp exists. Slack suits or coveralls are worn on jobs which require climbing or elevated positions.

Avoid cuts and scratches by wearing suitable gloves.

When piling material, place paper between each sheet.

Don't pile material with sharp edges protruding.

All parts of power saws are particularly dangerous, and material being sawed should be pushed past the saw with wooden blocks designed for this purpose.

Never let the hands come close to rapidly moving parts, such as router heads, band or circular saws, shears, and presses.

Always wear goggles or a face shield when cleaning with air, or when using any type of machine which may throw off particles of metal.

Always wear goggles when working on an overhead part.

When using files, care should be taken not to cut the hands on the edge of the material being filed.

Always clean, sharpen, and replace tools immediately after use.

Sharp tools should be covered when not in use.

Report all injuries, no matter how slight.

When driving out a rivet, be sure that there is no one in line with the rivet.

Keep fingers away from the jaws of a rivet squeeze.

Use a nozzle or blow gun on the hose when cleaning with air.

Warn all workmen near you before you start cleaning with air.

Don't disconnect air tools without shutting off the air pressure first.

Sweep up shot spilled from shot bags at once.

Don't take tool boxes on top of wings or center sections.

Don't leave parts on wings or stands where they may fall on men working below.

Replace corner pads and covers when through; think of the other fellow.

Don't start the motor with an extension drill until you are guiding the drill with one hand.

Care should be taken not to mar the work being drilled.

Don't run an extension drill near your hair, you may lose a portion of your hair in an instant.

Don't hold any small object while drilling it; clamp the work in a drill vise.

Don't operate any machine unless authorized to do so.

Don't use dull tools; use sharp ones and select the right tools for the job.

Don't use tools with split handles.

Don't use extension lights without guards; also, check for broken insulations.

Throw away wrenches that slip and don't fit.

Smooth up sharp edges on all jobs when possible.

Don't handle long, sharp-edged parts as if you were the only one present; look around first.

Don't reach into places where the mechanism can move until you take steps to assure safety.

Always report to First Aid with any scratch that draws blood; avoid infection.

Never drill a hole anywhere at any time without looking on the other side first.

Check all measurements carefully before drilling or cutting any material.

Always use a chuck key; it saves time, the drill, and the chuck.

Always see that the screw driver bit you are using fits the screw you are driving.

Never draw more stock than is required for the job you are doing.

Always report any condition that you think constitutes a safety hazard.

Ask all the questions pertaining to your job that you wish to, regardless of how foolish they may seem.

If you make a mistake, tell your supervisor at once.

Keep all tool boxes locked and in their proper places.

Keep your tool box clear of stock parts and in condition for inspection at all times.

Take time to pick up dropped screws, nuts, and bolts.

Put away light cords and air hoses when through using them and at the end of working hours.

Demand sufficient instruction to do your job right the first time. Don't guess.

Always replace guards when through working under them.

Don't leave an unfinished job without notifying your supervisor.

Screws and bolts must extend 1½ to 3 threads through stopnuts.

The standard practice with bolts or screws, whenever possible, is to have the head of the bolt or screw up, outboard, or to the front.

Do not use pliers of any kind on cannon plugs or breeze fittings; tighten by hand.

A screw driver sharpened like a knife will slip out of the screw slot. If ground too bluntly, it will not go to the bottom of the slot, hence it will raise a burr or break the screw. Since thousands of screws are used during the assembly of many airplanes, the importance of the proper use of properly conditioned screw drivers cannot be overestimated.

XI TECHNICAL DRAWINGS

It is becoming more and more important that the metal worker in aircraft construction be able to interpret engineering drawings rapidly and accurately. It is by means of this type of drawing that the workman can exactly duplicate the ideas of the designer. The workman usually receives a copy of the engineering drawings or parts of the original drawings representing the part which he is to build. These drawings are called "working drawings."

Most work is done from blueprints in the form of detailed drawings. There are a number of different ways in which these drawings are furnished to the workman. The blueprint is the most common. On blueprints the entire background is blue, and lines representing the drawing, dimensions, and figures appear in white.

There is also the brownprint which is not different from the blueprint except that the background is brown instead of blue. The lines on this drawing appear in white against a brown background. Blueprints and brownprints are made directly from the tracing, which is a duplication in ink of the drawing on translucent tracing cloth. Drawings are printed from the tracing cloth in much the same manner that a picture is printed from a negative on sensitive paper to make the finished photograph.

The blue-lined print is a print upon which the lines appear in blue against a white background. There are also brown-lined and black-lined prints.

By means of photography, photostats are obtained which may have white lines on a black background, which is the negative, or black lines on a white background, which is the positive. The latter method is used to enlarge or reduce drawings.

Pictorial or Oblique Views. Pictorial or oblique views picture the object in such a manner that three sides of the object are shown. Usually the sides which appear are the right side or face, the plan or top face, and the front face or front view. Such a view gives one the impression of seeing the whole object. On this type of drawing the lines are fore-shortened or drawn in perspective in order to make the object appear to have three dimensions. In engineering drawings the three faces which appear in the view are drawn as though seen at right angles to the face. For example, a rectangular block, as shown in Figure 81, viewed directly from the front appears as a rectangle. This is called the front view.

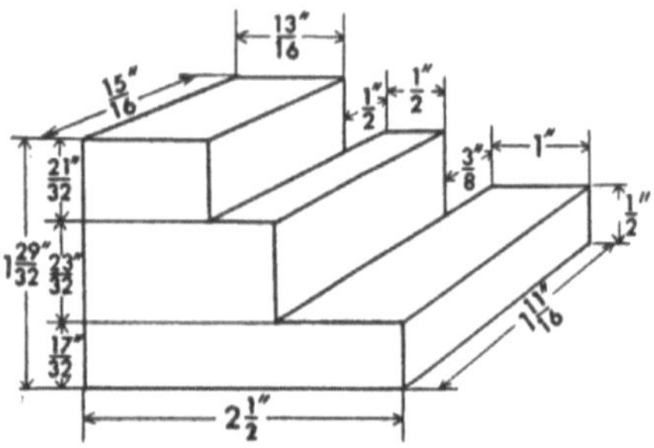

Fig. 80. Pictorial view of a solid with dimensions shown.

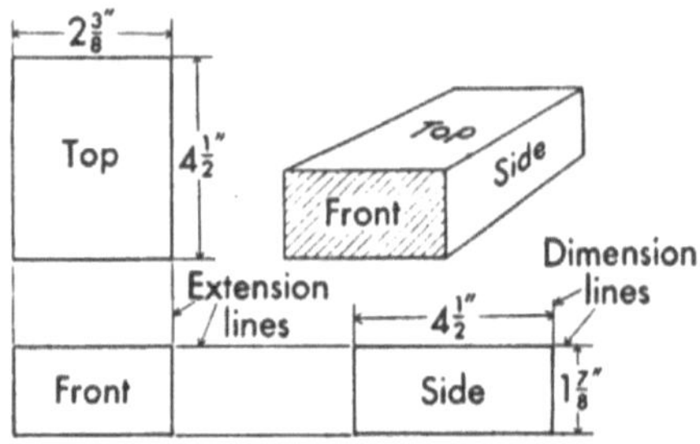

Fig. 81. Three-view drawing of a rectangular block of cast iron.

Viewed from the right side, another rectangle is seen which is drawn to the right of the front view and is called the side view. Directly above the front view is drawn the third rectangle which is called the top or plan view. In drawing these three views it is customary to place the front view in the lower lefthand part of the page, with the top view directly above it, and the right side view to the right of the front view. These three views are commonly called a "three-view drawing."

Blueprint Lines. Lines representing the edge of an object which can be seen are drawn heavily. These are called "visible boundary lines." To represent an edge which is hidden behind some other part of the object, fine dotted lines are used. These lines are called "invisible boundary lines." Sometimes a heavy boundary line on a view may represent both the visible boundary line and the invisible boundary line, when the invisible line is directly behind the visible boundary line.

The beginner usually has some trouble in getting a picture from these three flat views. With practice, however, he can, from these three views, get a mental picture of the complete object. As he gains experience, he can visualize a complicated object from the three views shown in this type of drawing.

The corresponding parts of each view are usually connected by thin, solid lines which are called "extension lines." Dimension lines show the distance between various parts of the object illustrated. Between these lines are placed a broken arrow and the dimensions in inches or feet. Dimension lines are illustrated on the drawings used in Figures 80 and 81.

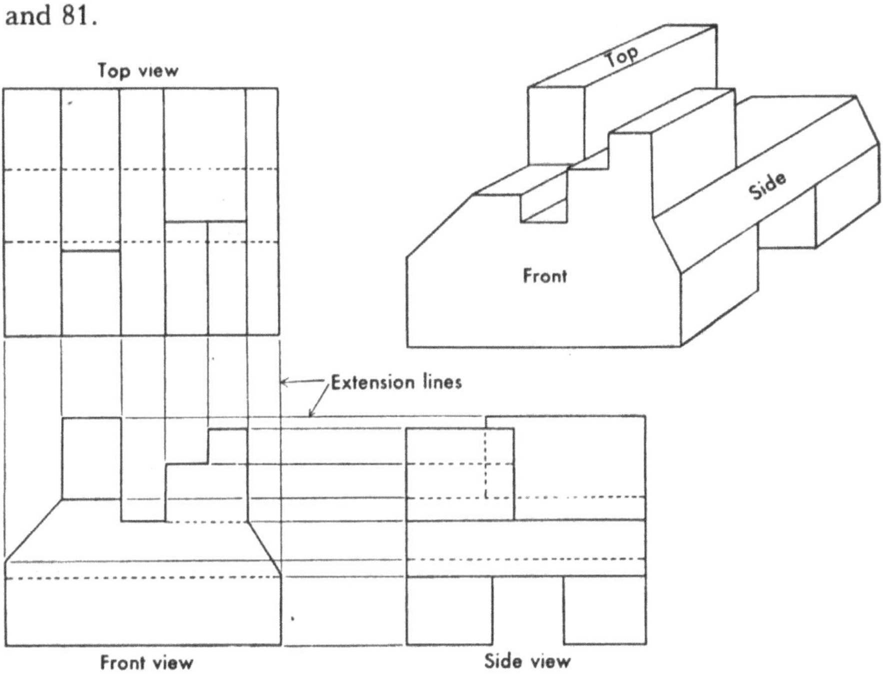

Fig. 82. Pictorial view and three-view drawing of the same object.

Sections. Sectional views are made by cutting through a part of the object and drawing the area which has been cut. These views are often used to show important details which cannot be otherwise shown on the drawing. Usually a line is made on the front, side, or top view,

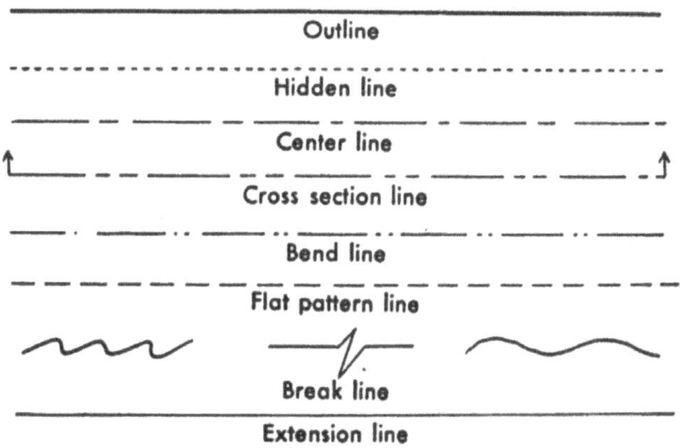

Fig. 83. Lines commonly used in technical drawings.

showing where the cut was made. Letters are used to indicate where the cut was made. On the sectional drawing, the same letters are used. A sectional view is usually marked with lines which represent the type

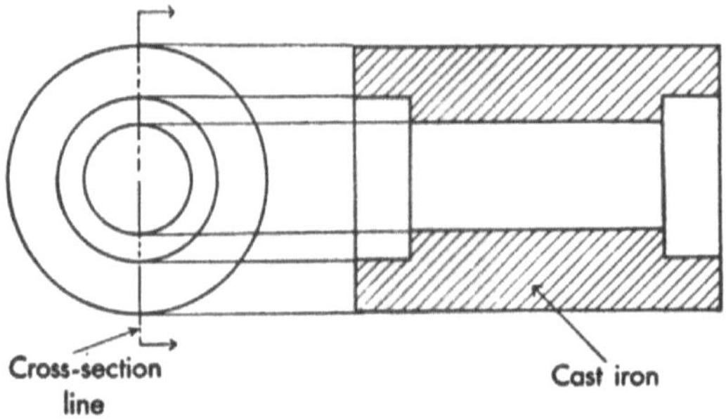

Fig. 84. End and cross section of a cast iron fitting.

of material cut, which may be wood, steel, brass, etc. These cross lines or markings are called "conventional cross-sectional markings."

A complete drawing or assembly drawing may represent either the complete structure which is being made, as for example an airplane; or a part of a complete structure, such as an airplane wing; or show how a smaller fitting is to be made. A detailed drawing shows only one small part in detail. Detailed drawings give all information needed to complete the part. Each detail is plainly shown on the drawing to enable the workman to complete the part. It would be almost impossible to build an airplane from a complete assembly drawing. Each part is shown in these detailed drawings and it is usually possible for one workman to complete all the work on the part shown on his detailed drawing.

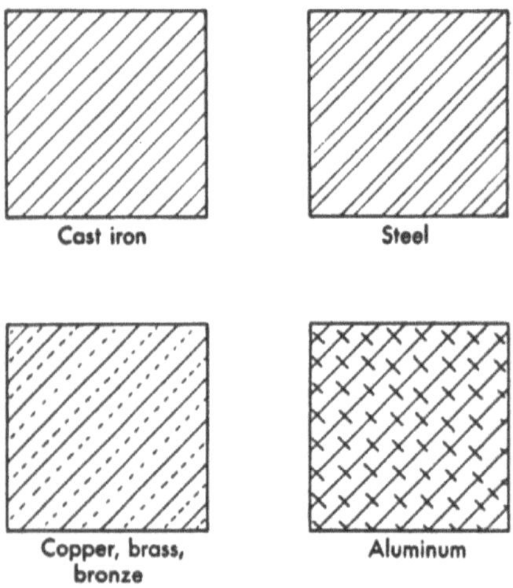

Fig. 85. Conventional cross-section markings used to indicate the kind of material cut.

Broken sections are used where it would be impossible to show a very long part on the drawing. For example, a spar may have its root and its tip very close together on the drawing, with the greater portion of the spar left out. This is made clear in Figure 86.

All blueprints and technical drawings follow the same pattern in order that any person familiar with this type of drawing will know exactly where to look to obtain needed information. In the lower right-hand corner of the drawing is usually placed a box with the title of the drawing in large letters. This title gives the name of the assembly and the name of the job. On a small detailed drawing, this title may simply give the name of the operation to be performed or the name of a single part to be made. Other information needed to clarify or make plain the title is often placed in this box, or in ruled-off portions of the box. The number of the drawing is placed to the right of the title. In the box will be found the name of the engineer who made the drawing or the company producing the part. The name of the draftsman may appear.

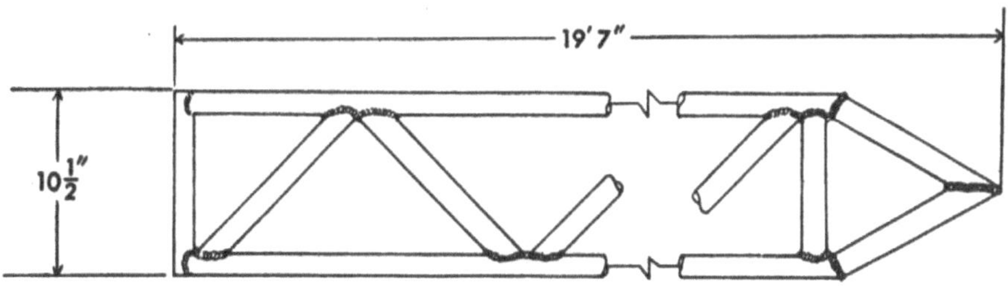

Fig. 86. Broken section of a welded tube spar.

If the drawing has been checked by another engineer or draftsman, the name of this person will also be placed in this box. The date of the drawing and the date approved are noted.

All materials to be used on the job are listed in a box near or above the title box. This box may have the word, noted, in it, which indicates that notes pertaining to the proper material are found elsewhere on the print. On blueprints the various parts are numbered. This enables the workman to get a detailed blueprint or drawing of that part.

Specifications are directions which accompany drawings describing the kinds of material to be used. The driving of an extra nail or the placing of an extra screw might weaken instead of strengthen the part.

Each workman should have a good understanding of blueprints and drawings. A person learns to read a blueprint much as he learns to read the printed page. A blueprint which looks complicated to the beginner is as readable to a person familiar with the symbols and lines as is the printed matter in a simple school book.

XII BURRING AND FILING

When sheet metal is sheared, punched, sawed, or drilled, the edge of the metal is usually left in a roughened condition. The sharp points left by the cutting tool are called "burrs." Tiny cracks may also appear on the cut edge. The removal of these burrs and cracks is called "burring." Proper burring leaves the edge of the metal smooth and even.

Burring is done for several reasons. Cracks too small to be seen by the naked eye, across the edge of a piece of sheet metal may cause minute cracks to develop while the metal is being formed. These minute cracks will become enlarged and cause failure of the part in

Fig. 87. Burring stamped parts that are to be formed. (Courtesy Reynolds Metals Company)

service. Burrs along the edge of parts may prevent a perfect fit. Burrs along the edge of pieces of sheet metal are almost sure to scratch and damage other sheets with which they come in contact. Serious cuts and scratches may be caused by burrs. Figure 88 shows the appearance of a magnified edge of a piece of sheet metal which has

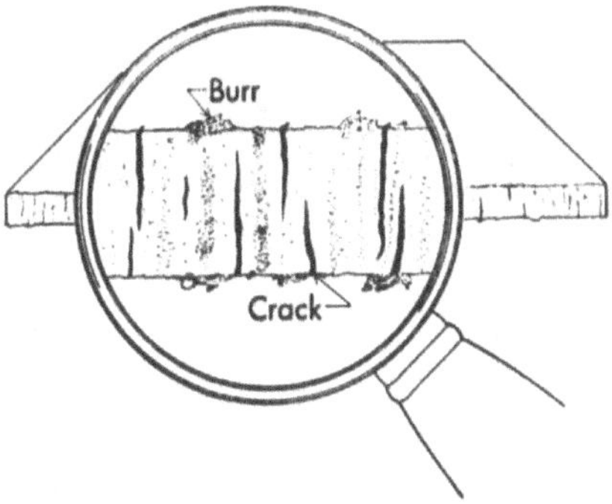

Fig. 88. Magnified edge of a piece of sheet metal that has been cut.

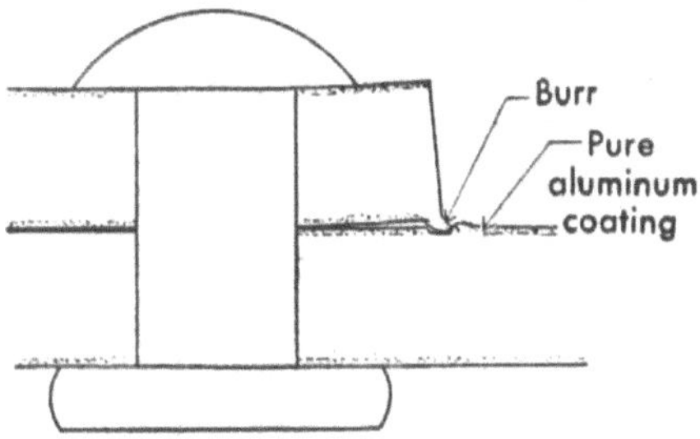

Fig. 89. Damage may be caused by burrs left on the edge of riveted sheets.

been cut. However, before burring a part, be sure that burring is necessary. Many production parts do not need to be burred.

In burring, not only the burrs on each edge of the sheet must be removed, but also enough of the edge of the sheet itself to remove any cracks which might have been formed. Any edge which is to be bent or stretched must be burred. Edges that have been sawed are rough

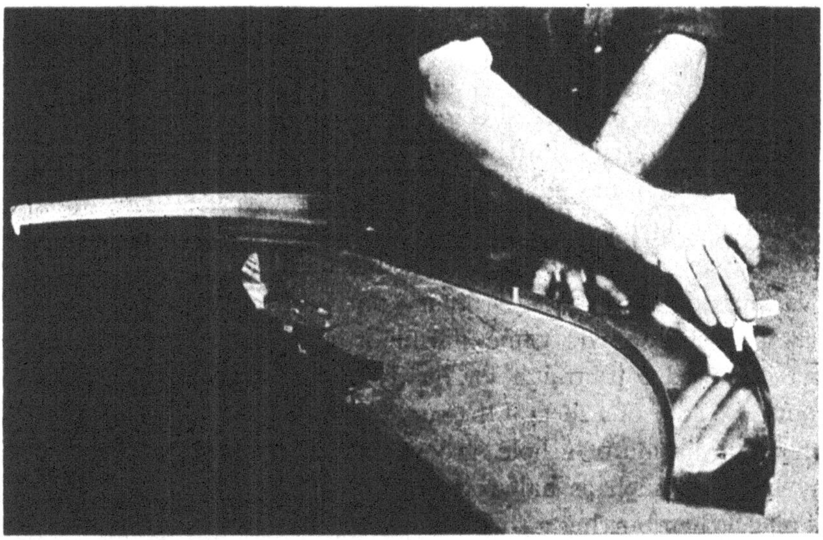

Fig. 90. Hand-burring a formed part. (Courtesy The Glenn L. Martin Company)

and ragged, but sawing does not usually cause cracks to appear in the edge of the sheet. Cracks may, however, start at the base of the small grooves left by the saw teeth. Sawed material should have both the burrs removed and the edges smoothed. On drilled holes, it is usually only necessary to remove the rough burrs around the edge of the hole.

Fig. 91. Grinding wheels used for burring. (Courtesy Reynolds Metals Company)

Routed material should be burred and smoothed. Sheared material usually develops small cracks and should be burred and smoothed. Sawed, punched, or sheared edges are more likely to fail when forming than are routed edges. Parts which have small tolerances must be carefully burred and the edges smoothed. Edges of sheet that are to be riveted must have all burrs removed.

Burring may be done by several different methods and by the use of various kinds of tools. Ordinary flat files may be used to remove burrs if carefully used. A fine, flat file is used to smooth the edge of the sheet if the edge is straight or nearly straight. A file may be used to smooth the edge of a sheet if the curve is convex. If the curve is concave, such as the inside of a hole, groove, or corner, files of various shapes may be used, such as half round, round, pippen, square, triangular, or knife. When burring a straight edge, the file should be pushed slantwise along the edge and it should be pushed away from the workman.

When filing soft metals such as aluminum or Dural, allowing the file to drag lightly on the return stroke helps to keep it clean. In using a round or half-round file when smoothing a curve, the file should be rolled slightly in a clockwise direction as the stroke is made. This rolling motion allows the file to cut better and distributes the wear over all

Fig. 92. Removing burrs on a vertical sander. (Courtesy The Glenn L. Martin Company)

of the teeth. Sharp scrapers, such as hook scrapers or bearing scrapers, may be used for removing burrs. A V-edge scraper or a scraper containing small steel wheels similar to those in a kitchen knife sharpener may be used. Oversized drills, hand countersinks, or special burring tools are used to remove burrs from holes. In using rotary files or grinding wheels for burring, extreme care must be taken to remove only the desired amount of material, as power machinery cuts rapidly and damage to the part may result. In production work, rotary sanders, steel-wire brushes, abrasive wheels, and steel rolls are used for burring and finishing the edges of cut sheet. Steel wool wrapped around a shaft attached to a motor may be used to burr and finish the edges of stamped or cut parts.

Filing is an operation which is used frequently in the working of sheet metal. Files are of many types, and it is necessary to pick the right file for the job. All files should be equipped with a wooden handle while in use. The part of the file which fits into the handle is called the "tang." Figure 94 shows the various parts of a file. The file cuts in much the same manner as a saw. On a single-cut file, the teeth extend

Fig. 93. Removing burrs on a horizontal sander. (Courtesy The Glenn L. Martin Company)

across the face of the file. On a double-cut file, the teeth are cut across the face in two directions. A bastard file may be either single or double cut and has coarse teeth. A vixen file has curved teeth. A mill file has fine teeth and is a good all-purpose file for fine finishing. A Swiss file comes in various shapes and has fine teeth. Files also come in other shapes, such as flat, round, half-round, pippen (which is an oval shape) square, triangular, or knife edged. They should be kept clean with a

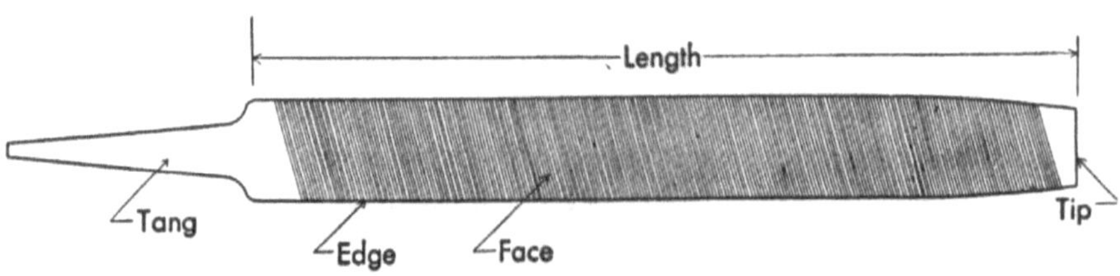

Fig. 94. The various parts of a file.

file card which has a wire brush on one side and a bristle brush on the other. The wire brush is used to loosen the particles in the file teeth and the bristle to brush them away. Files should be wrapped in a cloth and separated from each other when put away. When using a file, usually the tip is held in the left hand and the handle of the file in the right hand. The strokes should be even, and the file should not be allowed to rock on the work if the surface is to be straight and level. Filing requires considerable practice in order to be done properly.

XIII LAYOUT

The technical problems involved in layout are those of the draftsman and engineer. However, it is necessary that the sheet-metal worker be able to use the fundamental layout tools. The workman should be able to square material to be cut, locate rivet holes, lay out material, and develop patterns, in order that he may cut the material to the required size. He should be able to use protractors, compasses, dividers, straight edges, trammel gauges, scribers, squares, and measuring instruments.

Layout, when it concerns the aircraft sheet-metal worker, consists largely of transferring information contained on a blueprint to a paper pattern or to the material with which he is to work. All of the information necessary to complete a job is usually found on a blueprint, although the drawing on the blueprint is not usually full size. It is necessary that the

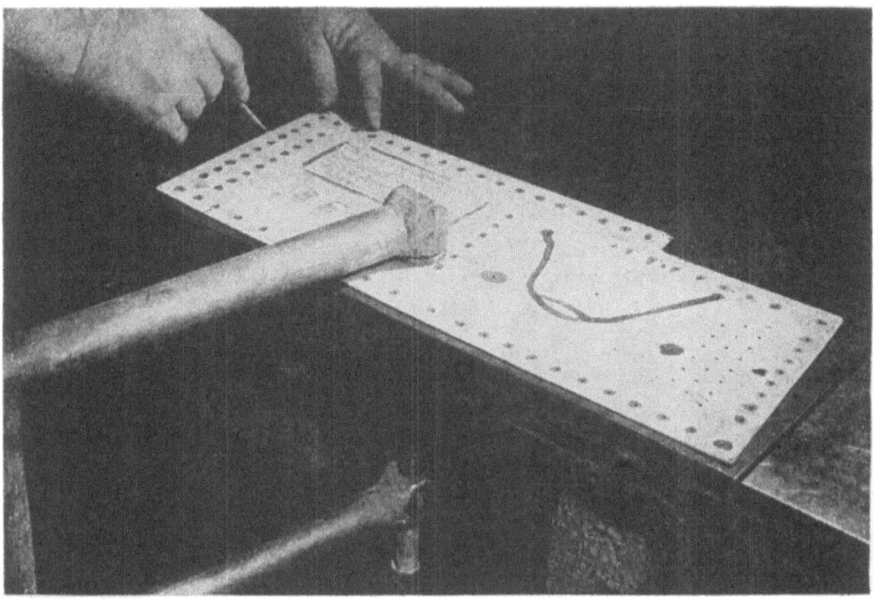

Fig. 95. Using a template as a pattern to lay out a part. (Courtesy Douglas Aircraft Company, Inc.)

workman be able to lay out a full-scale drawing. This layout may be made on the material itself, but if the job is in any way complicated, the layout should first be made on a sheet of heavy paper which may then be cut to the proper size and used as a pattern by which to cut the metal. When making a layout, the workman should use extreme care to check the pattern with the blueprint in every detail. The paper

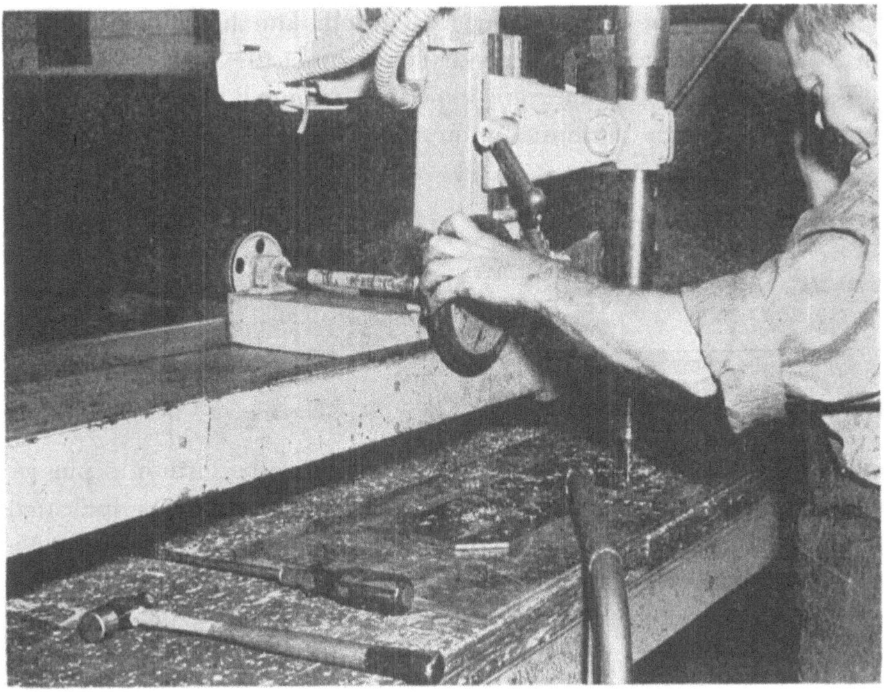

Fig. 96. Using a template to locate holes being drilled. (Courtesy Douglas Aircraft Company, Inc.)

pattern should be fastened to the material with adhesive tape, and the outline drawn on with a soft, sharp pencil. Care should be taken to hold the pencil in such a manner that the line closely follows the edge of the pattern. Most aircraft metals do not permit the use of a metal scriber. The scratches weaken the material and may lead to corrosion and the development of cracks. Holes to be drilled may be located by a center punch or prick punch, taking care not to damage the material. When the use of a punch is not allowed, the centers of the holes may be located by crossed pencil marks.

When a considerable number of duplicate pieces are to be made, templates should be prepared. A template is an exact pattern made of mild sheet steel or other material which may be coated with a dye in order that any markings may be readily seen. This coating also protects the metal from corrosion. Templates may be in the form of flat patterns or in other shapes which will test the accuracy with which a part has been made. They may be made to clamp to the part being formed while trimming with a router or nibbler, or for the accurate drilling of holes. A template is a precision instrument and should be handled with care to prevent damage. The template should have marked on it all of the information necessary for forming the part.

Tolerance and allowance are sometimes confused. Tolerance is the amount of variation permitted in the size of a part. Tolerance is usually

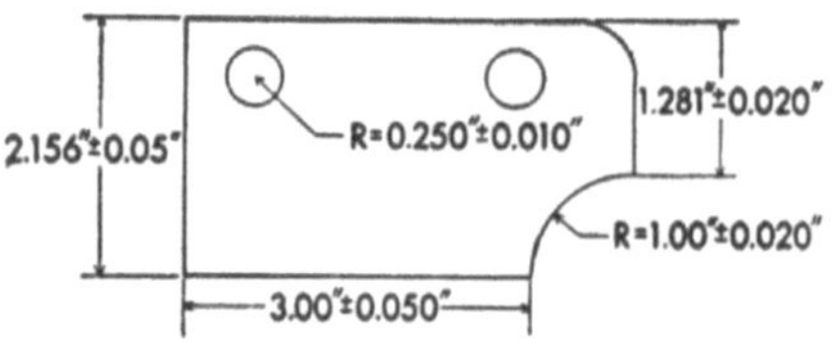

Fig. 97. Tolerance.

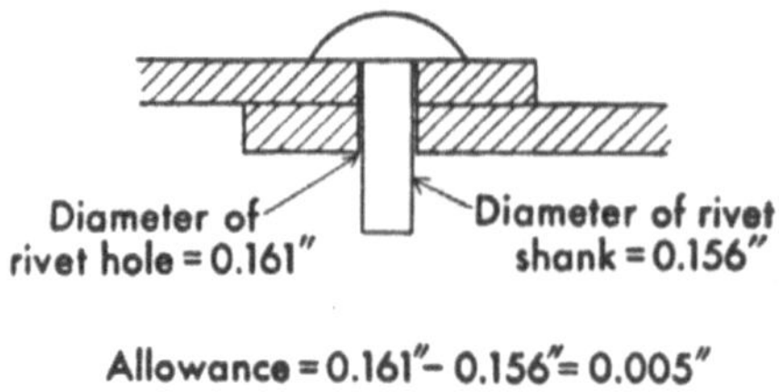

Fig. 98. Allowance.

indicated on the blueprint or on the template in the form of a plus or minus fraction. For example, the angle of a flange might be indicated as 45° ± ½°. This would mean that the workman must bend the angle to within one-half of a degree of the required angle. The length of a part might be indicated on the drawing as 25½ in. ± 1/32 in. Most airplane parts have a tolerance of not more than 1/32 in.

Allowance is the difference in dimensions between two parts to allow the proper fit. For example, the piston of an airplane engine does not have the same diameter as the cylinder, because there is an allowance for the expansion of the piston. A rivet hole is not exactly the same size as the rivet as an allowance is made so that the rivet may be put easily into place. Two sheets that join end to end do not fit exactly because allowance is made for expansions and tolerance.

The following are typical abbreviations which are used on templates by the Douglas Aircraft Company, Inc.

A Apply.

D Drill.

C Contour.

T Trim or (at end of abbreviation) template.

ADT Apply drill template, used to locate holes to be drilled.

ADCT Apply drill contour template, used as an ADT, but with some contour such as an angle or bend.

ATT Apply trim template, used to trim a part to a desired size.

ATDT Apply trim and drill template, used as both an ADT and an ATT.

ATCT Apply trim contour template, used to trim, form, or check a part with a contour.

ATDCT Apply trim and drill contour template, used as a combination of ADCT and ATCT.

CT Contour template, used to form a part to a contour, and/or check the correctness of that part.

FBT Form-block template, used as a pattern for making a form block.

GDT Gang-drill template, used to locate holes to be gang-drilled.

JDT Jig-drill template, used to locate holes to be drilled in parts on an assembly jig.

LT Layout template, used as a flat pattern to which a sheet must be cut, drilled, or formed.

NT Nibbler template, used for making a part on the nibbling machine.

PBT Pierce blank template, used for piercing and blanking a part.

PT Profile template, used for making a part on the profiling machine.

RST Rough-saw template, used for cutting the sheet stock for a part to be fabricated in a later operation.

RB Router block, used for making parts on a router machine.

SST Stock-size template, used for cutting the flat outline for a part to be formed.

JOG Joggle: an offset made to cover the abutting end of an adjacent part. Unless stamped otherwise, all joggles are up.

BL Bend line: a line where the surface begins to bend, or at which the surface begins to go straight again after a bend.

BPH Brake pin hole: the hole for a pin which keeps the template and part in place on a power brake.

CL Closed, as in 3° CL, which indicates the degree of a bend over toward the surface from which the bend starts.

2°CL Bevel: the degree by which an angle is greater or less than a right angle.

D or **DIA** Diameter.

DIR RUN Direction of run: an indication used as a guide to cutting stock, so that the blank may be run in correct direction through a fabricating operation such as a punch press.

DN Down: the direction in which a flange or joggle is to be made, as JOG DN.

FLG Flange: a bend at the edge of sheet stock, made at some specified angle to the surface. The width of flange and radius of bend are indicated thus: FLG DN 3/16″ R 3/32. If no angle is specified, such as 5° OP, the bend is at a right angle.

LTG HOLE Lightening hole: a hole made in a member in order to reduce weight or to stiffen it by means of a flange around the hole.

ML Mold line: the intersection of the outside surfaces of a sheet of metal that is bent. It is an imaginary line.

OC On center: the center-to-center distance between holes, as "pierce 1" OC.

OP Open, as in 5° OP, indicates the degree of a bend which opens out away from the surface from which it starts.

OPP Opposite: this abbreviation is always a part of the drawing number. It means that all jogs, flanges, etc., are to be made in the opposite direction from indications on the template.

OS up Other side up: indicates that the other side of the template must be up when applied to the part.

PH Pin hole: a hole made to receive a locating pin.

R Radius: the term may refer to the radius of a bend or flange, or to the radius of bend relief holes.

REF LINE Reference line: an arbitrary line used only as a means of locating points or lines of a part.

RELIEF Relief: removing metal from the place where two bends or flanges meet in order to prevent wrinkling made by excess metal.

RPH Router pin hole: a pin hole for a locating pin on a router block.

SB Setback: the difference in the amount of metal needed to make a bend and the amount of metal that bend would have needed if it had been made at a completed angle instead of around an arc.

SHN Shown, which means that the part is to be made exactly as shown on the template. Reverse of OPP.

STA Station: the cross section of a large contoured part, such as a wing or fuselage.

TAB Tab: a section of a template useful only to stiffen a template (in which case it is sometimes called a brace); or to make the surface larger, as to accommodate a locating pin; or, in the case of a very small template, to carry all needed directions about its use and application. It is never reproduced on a part.

TAPER JOG Taper jog: a joggle which goes off at an angle from the main surface after the offset, instead of parallel to it.

TH Tooling hole: a hole made to meet the needs of tooling rather than of fabrication, as for example, to coordinate parts on an assembly jig.

TOA Trim on assembly: stamped-on edges which are to be trimmed to fit.

TPH Tooling pin hole: a TH which is also a PH. Tooling pin holes must fit over the corresponding pins without being forced on, as with a hammer.

TS up This side up: means that the template is to be applied with the indicated side up.

UP Up: indicates the direction of a bend, flange, or bead.

Mock-up is a term used to describe a model which is built full size. This model may be designed to represent the fuselage of an airplane, a complete wing, or any other complete part. The mock-up is used to determine whether or not installed parts will fit. The mock-up may be used as a template to test the accuracy of the various parts, or it may be used as a die.

XIV BEND ALLOWANCE

Most work done on sheet metal has to do with changing its shape. This shaping is primarily bending. The bend may be slight or severe, and there may be a number of bends of different kinds on a single piece of material. In order to avoid wasting material, it is important that the sheet-metal worker be able to determine exactly how much of the sheet will be required to complete the job. In forming a flange, an unskilled workman might take a piece of metal somewhat larger than the piece required and bend it to the required angle, then trim off the excess material and finish the edges to exact size. This operation would not only waste material, but the waste in time required to trim and finish the part might increase its cost many times. One large aircraft factory states that the use of bend allowance in sheet-metal development has enabled the aircraft industry to save billions of dollars every year in man-hours and material.

Bend allowance, once thoroughly understood, becomes simple enough for the workman to use it easily and intelligently. A thorough understanding of the meaning of bend allowance should be had before attempting to work problems in bend allowance or in layout where bend allowance is involved. It is not enough to be able to use tables and formulas blindly.

Bend allowance is the amount of material which must be allowed to form the bend. This allowance is the amount of material actually needed to form the curved portion. For example, it may be necessary to bend a sheet of material to form an angle of 90°. On extruded parts, angles may be of any shape, having bends of small radii or sharp corners, as shown in Figure 99. Most sheet metal, if bent to form a sharp corner, would crack or become so weakened that it would fail in service. When sheet metal is bent, the angle formed should not have a sharp edge. The vertex of the angle formed by the sides is not sharp, but follows the arc of a circle. The radius of this arc is called the "bend radius."

The kind of metal and its thickness determines the bend radius. The bend radius is the radius of the arc forming the inner side of the bend, as shown in Figure 100. Bending material both stretches and shrinks the metal. The metal on the inside of the bend shrinks, or is squeezed together, while the metal on the outside of the bend is stretched. The

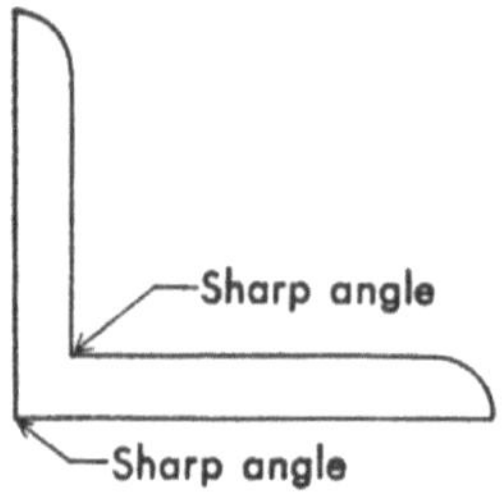

Fig. 99. An extruded part — short bend and sharp angle.

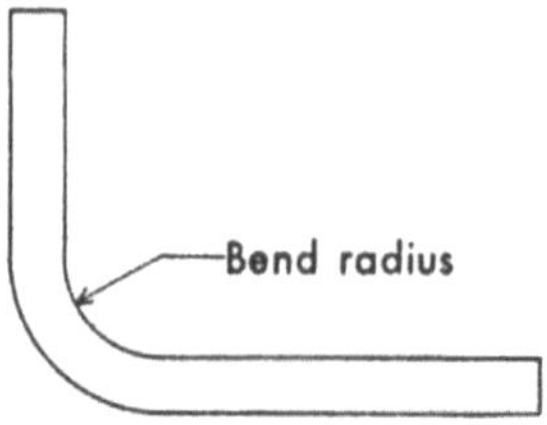

Fig. 100. Bend radius.

line along which no stretching or shrinking takes place is called the "neutral line." This neutral line is not exactly halfway between the surfaces of the metal being bent. It is nearer the inside of the bend, being 44.5 per cent of the thickness of the metal from the inner surface, as shown in Figure 101. In laying out sheet metal to form an angle, the

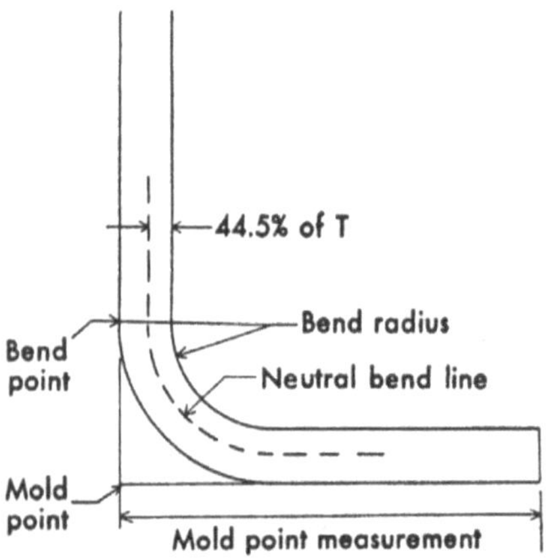

Fig. 101. Neutral bend line.

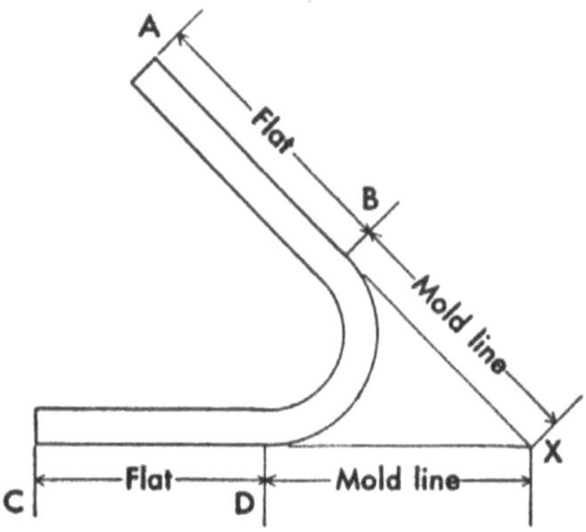

Fig. 102. Flats and mold line.

line forming the outside of the angle is extended until it meets the line forming the outside of the other side of the angle. These lines form the "flats" of the angle until they reach a point where the bend begins, as shown in Figure 102. The line from this point to the vertex of the angle thus formed is called the "mold line." The distance from the point of bend to the vertex of this angle is called "setback." In an angle of 90°, this setback is equal to the bend radius plus the thickness of the metal. This statement is written as $R + T$.

From Figure 102 it is readily seen that the amount of material required to form the angle is equal to the distance from *A* to *B*, this being the flat of one side, plus the distance from *C* to *D*, which is the flat of the other side, plus the material forming the bend itself. The length of *AX* is the flat of one side plus the mold line, the mold line being the distance from *B* to *X*. The distance from *B* to *X* is equal to the length of the flat on the other side plus the mold line from *C* to *X*. These are the distances which are usually given in laying out the angle. If the distance, *AX*, is equal to 1½ in., and the distance, *BX*, is equal to 2 in., it might be easy to assume that the overall length, or developed flat pattern for the material needed, would be the sum of these two measurements, or 3½ in. However, if a strip of metal 3½ in. long were bent to form the angle, it would be found that the sides of the angle exceeded the desired measurements given on the drawing. This is because the metal does not extend to the vertex of the angle but cuts across the corner, requiring a smaller amount of metal than that required to form a true angle.

The circumference of a circle is less than the length of the sides of a square having the same radius. The amount of metal required to form the bend itself is always less than the sum of the two mold lines forming the angle, as shown in Figure 102. The amount of material necessary to form the bend may be figured by finding the portion of the circumference of a circle equal in degrees to the angle formed by the sheet metal. The radius of this circle would be equal to the bend radius plus 44.5 per cent of the thickness of the metal. For example, if a 90° bend were to be formed in material 0.032 inch thick having a bend radius of ⅜ in., the amount of material necessary to form the bend could be found as follows: The circumference of the neutral circle may be found by using the formula $2\pi R$; π is equal to 3.1416. To find the radius of the neutral circle, multiply the thickness of the metal (0.032 inch) by 0.445 (the decimal equivalent of 44.5 per cent) and add this answer to the bend radius, which is 0.375 (decimal equivalent of ⅜); 44.5 per cent of the thickness of the metal is 0.014; 0.375 plus 0.014 gives 0.389, which is the radius of the neutral circle; 0.389 times 2 times π is 2.444, which is the circumference of the neutral circle to the nearest one-thousandth of an inch. Since only ¼ of the entire circumference is needed to form the bend, this number is divided by 4, which is 0.611 and is the amount of metal needed to form the bend itself.

If the angle had been formed through 45°, the amount of material needed to form the bend would be ½ that above. To form a complete circle, the amount of material needed would have been equal to the circumference of the neutral circle, or 2.444. To form an angle of 1°, the circumference of the neutral circle, which is 2.44416, is divided by 360 (the number of degrees in a circumference), giving 0.00679. This is the amount of material needed for each degree of bend when the radius is ⅜ in. and the thickness of the material is 0.032.

To save the above figuring each time an angle is laid out, tables have been prepared showing the amount of material or bend allowance for each degree of bend, or the amount of material needed for a 90° bend in materials of various thicknesses and different bend radii.

In the lefthand column of Table IV, look up ⅜ in., which is the bend radius; then find, in the top row of figures, the thickness of the material, 0.032. Where these two columns cross each other, we come to the figure, 0.00679, which is the amount of material found necessary in the preceding problem to form a one-degree bend.

When the sheet-metal worker is thoroughly familiar with this procedure, he may approach the next step in figuring bend allowance which is finding setback, so that he may determine accurately the length of the flats forming each side of the angle. The setback is the distance from the vertex of the angle, formed by the mold lines, to the point where the bend begins on each side of the angle.

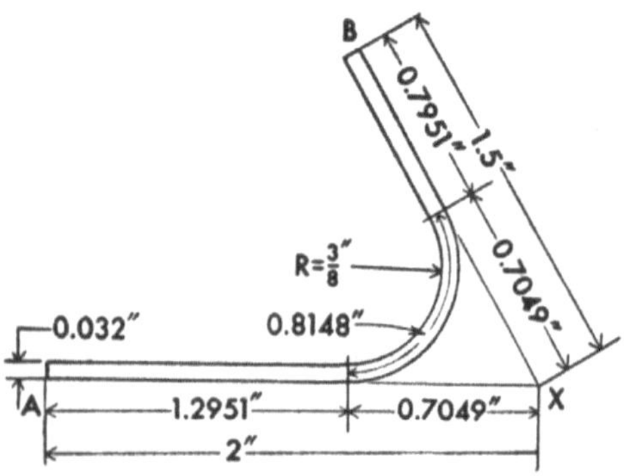

Fig. 103. Empirical formula — 60 degrees.

To find the setback an empirical formula is used. An empirical formula is one which has been worked out by experiment. The setback is found by adding the thickness to the bend radius ($T + R$) and multiplying this number by the cotangent of one-half the angle formed by the sides, as shown in Figure 103. For example, the angle formed by the sides in Figure 103 is 60°. From a table of natural cotangents, it is found that the cotangent of 30°, which is one-half the angle formed by the sides, is 1.7320. If the bend radius is ⅜ in., and the thickness of the material is 0.032 in., the formula becomes:

$$(T + R) \cot \frac{\text{angle}}{2} = (0.032 + 0.375)\ 1.7320 = 0.7049$$

This number represents the setback which must be subtracted from the distance from A to X and from B to X to determine the length of the flat on each side of the angle. Since the distance from A to X is 2 in. and from B to X is 1½ in., the flat, A to C, is equal to $2 - 0.7049 = 1.2951$. In a similar manner, the flat forming B to D is equal to $1.5 - 0.7049 = 0.7951$. It can now be seen that the amount of material needed to form the given angle is 1.2951, which is the distance from A to C plus 0.7951, which is the distance from B to D plus the bend allowance. Since the angle is equal to 60°, we find from the table for 1° bends the amount of material needed when the thickness is 0.032 and the bend radius is ⅜ in. to be 0.00679. This number multiplied by 120, since the metal is bent through 120° to form a 60° angle, is 0.8148, which is the amount of material needed to form the bend. Adding 1.2951, 0.7951, and 0.8148, it is found that the required amount of material to form the given angle is 2.9050. This is called the developed flat pattern, as shown in Figure 104.

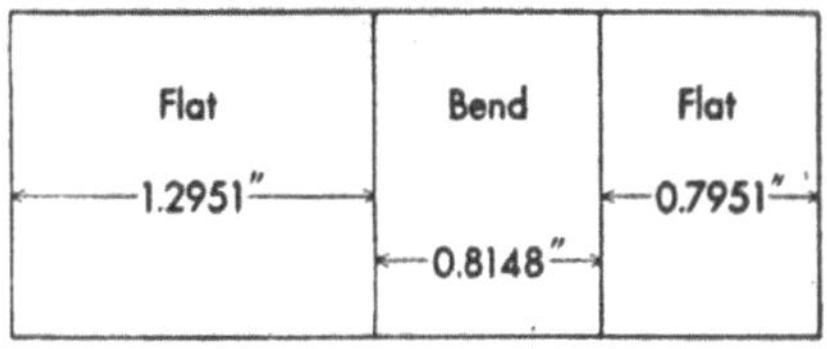

Fig. 104. The developed flat pattern.

Another empirical formula, instead of using the cotangent of one-half the angle formed between the sides, uses the tangent (tan) of one-half the angle through which the metal has been bent. The values obtained would be the same because each angle is the supplement of its other. The cotangent of an angle is equal to the tangent of the supplementary angle. Supplementary angles are two angles whose sum is equal to 180°.

A practical shop bend allowance formula, as given below, is often used.

Bend allowance formula:

$$[(.01743 \times R) + (.0078 \times T)] \times \text{no. degrees of bend}$$

The procedure for using the above formula is as follows.

1. When the inside dimensions of a bend are given, subtract one radius for each dimension.

TABLE IV. TABLE FOR FINDING AMOUNT OF MATERIAL FOR EACH DEGREE OF BEND

R \ T	.010	.012	.015	.018	.020	.022	.025	.028	(1/32) .031	.032	.035	.040	.045	.049	.050	.051	.057	.062	.064	.065
1/64	.00035	.00037	.00039	.00041	.00043	.00044	.00047	.00049	.00051	.00052	.00055	.00058	.00062	.00065	.00066	.00067	.00072	.00076	.00077	.00078
1/32	.00062	.00064	.00066	.00068	.00070	.00072	.00074	.00076	.00079	.00079	.00082	.00086	.00090	.00093	.00093	.00094	.00099	.00103	.00104	.00105
3/64	.00089	.00091	.00093	.00096	.00097	.00099	.00101	.00103	.00106	.00107	.00109	.00113	.00117	.00120	.00121	.00121	.00126	.00130	.00132	.00132
1/16	.00117	.00118	.00121	.00123	.00125	.00126	.00128	.00131	.00133	.00134	.00136	.00140	.00144	.00147	.00148	.00149	.00153	.00157	.00159	.00160
5/64	.00144	.00145	.00148	.00150	.00152	.00153	.00156	.00158	.00160	.00161	.00163	.00167	.00171	.00174	.00175	.00176	.00181	.00184	.00186	.00187
3/32	.00171	.00173	.00175	.00177	.00179	.00181	.00183	.00185	.00188	.00188	.00191	.00195	.00199	.00202	.00202	.00203	.00208	.00212	.00213	.00214
7/64	.00198	.00200	.00202	.00204	.00206	.00208	.00210	.00212	.00215	.00216	.00218	.00222	.00226	.00229	.00230	.00230	.00235	.00239	.00241	.00241
1/8	.00226	.00227	.00230	.00232	.00233	.00235	.00237	.00239	.00242	.00243	.00245	.00249	.00253	.00256	.00257	.00258	.00262	.00266	.00268	.00269
9/64	.00253	.00254	.00257	.00259	.00260	.00262	.00265	.00267	.00269	.00270	.00272	.00276	.00280	.00283	.00284	.00285	.00290	.00293	.00295	.00296
5/32	.00280	.00282	.00284	.00286	.00287	.00290	.00292	.00294	.00296	.00297	.00300	.00304	.00307	.00311	.00311	.00312	.00317	.00321	.00322	.00323
11/64	.00307	.00309	.00311	.00313	.00314	.00317	.00319	.00321	.00324	.00324	.00327	.00331	.00335	.00338	.00339	.00339	.00344	.00348	.00349	.00350
3/16	.00335	.00336	.00339	.00341	.00342	.00344	.00346	.00348	.00351	.00352	.00354	.00358	.00362	.00365	.00366	.00367	.00371	.00375	.00377	.00378
13/64	.00362	.00363	.00366	.00368	.00370	.00371	.00374	.00376	.00378	.00379	.00381	.00385	.00389	.00392	.00393	.00394	.00398	.00402	.00404	.00405
7/32	.00389	.00391	.00393	.00395	.00396	.00398	.00401	.00403	.00405	.00406	.00409	.00412	.00416	.00419	.00420	.00421	.00426	.00430	.00431	.00432
15/64	.00416	.00418	.00420	.00422	.00423	.00426	.00428	.00430	.00433	.00433	.00436	.00440	.00444	.00447	.00448	.00448	.00453	.00457	.00458	.00459
1/4	.00443	.00445	.00447	.00450	.00451	.00453	.00455	.00457	.00460	.00461	.00463	.00467	.00471	.00474	.00475	.00475	.00480	.00484	.00486	.00486
9/32	.00498	.00500	.00502	.00504	.00505	.00507	.00510	.00512	.00514	.00515	.00517	.00521	.00525	.00528	.00529	.00530	.00535	.00539	.00540	.00541
5/16	.00552	.00554	.00556	.00558	.00559	.00562	.00564	.00566	.00569	.00570	.00572	.00576	.00580	.00583	.00584	.00584	.00589	.00593	.00595	.00595
11/32	.00607	.00608	.00611	.00613	.00614	.00616	.00619	.00621	.00623	.00624	.00626	.00630	.00634	.00637	.00638	.00639	.00644	.00647	.00649	.00650
3/8	.00661	.00663	.00665	.00667	.00668	.00671	.00673	.00675	.00678	.00679	.00681	.00685	.00689	.00692	.00693	.00693	.00698	.00702	.00704	.00704
13/32	.00716	.00717	.00720	.00722	.00723	.00725	.00728	.00730	.00732	.00733	.00735	.00739	.00743	.00746	.00747	.00748	.00752	.00756	.00758	.00759
7/16	.00770	.00772	.00774	.00776	.00777	.00780	.00782	.00784	.00787	.00788	.00790	.00794	.00798	.00801	.00802	.00802	.00807	.00811	.00812	.00813
15/32	.00825	.00826	.00829	.00831	.00832	.00834	.00837	.00839	.00841	.00842	.00844	.00848	.00852	.00855	.00856	.00857	.00861	.00865	.00867	.00868
1/2	.00879	.00881	.00883	.00885	.00886	.00889	.00891	.00893	.00896	.00896	.00899	.00903	.00907	.00910	.00911	.00911	.00916	.00920	.00921	.00922
17/32	.00934	.00935	.00938	.00940	.00941	.00943	.00945	.00947	.00950	.00951	.00953	.00957	.00961	.00964	.00965	.00966	.00970	.00974	.00976	.00977
9/16	.00988	.00990	.00992	.00994	.00995	.00998	.01000	.01002	.01005	.01005	.01008	.01012	.01016	.01019	.01019	.01020	.01025	.01029	.01030	.01031
19/32	.01043	.01044	.01047	.01049	.01050	.01052	.01054	.01057	.01059	.01060	.01062	.01066	.01070	.01073	.01074	.01075	.01079	.01083	.01085	.01086

Bend allowance chart, for any degree angle of bend. (Courtesy of The Glenn L. Martin Company)

TABLE V. TABLE TO FIND DEVELOPED LENGTH OF 90° BENDS

T \ R′	1/32	3/64	1/16	5/64	3/32	7/64	1/8	9/64	5/32	11/64	3/16	13/64	7/32	15/64	1/4	9/32	5/16	11/32	3/8	7/16	1/2
.330																	.499	.512	.525	.549	.579
.284																.466	.479	.492	.506	.529	.557
.259															.422	.435	.449	.462	.476	.499	.529
.238														.390	.396	.410	.423	.437	.450	.473	.504
.220													.361	.368	.374	.388	.401	.415	.428	.451	.482
.203												.333	.340	.347	.354	.367	.381	.394	.408	.431	.461
.180											.299	.306	.313	.319	.326	.339	.353	.366	.379	.403	.433
.165										.274	.281	.288	.294	.301	.308	.312	.335	.348	.361	.385	.415
.148									.247	.254	.260	.267	.273	.280	.287	.300	.314	.327	.340	.364	.394
.134								.223	.230	.237	.243	.250	.257	.263	.270	.283	.297	.310	.324	.347	.377
.120							.199	.206	.213	.220	.226	.233	.240	.246	.253	.266	.280	.293	.307	.330	.360
.109						.179	.186	.193	.199	.206	.213	.220	.226	.233	.240	.253	.263	.276	.243	.317	.347
.095					.156	.162	.169	.176	.182	.189	.200	.203	.204	.216	.223	.236	.250	.263	.276	.300	.330
.083				.135	.141	.148	.154	.161	.168	.175	.181	.188	.195	.201	.208	.222	.235	.248	.262	.285	.315
.072			.144	.121	.128	.134	.141	.148	.155	.161	.172	.175	.181	.188	.195	.208	.222	.235	.248	.272	.302
.065		.100	.106	.113	.119	.126	.133	.139	.146	.153	.164	.166	.173	.180	.186	.200	.213	.226	.240	.263	.293
.058	.084	.092	.098	.104	.111	.117	.124	.131	.138	.144	.155	.158	.164	.171	.178	.191	.205	.218	.231	.255	.285
.049	.073	.081	.087	.093	.100	.107	.113	.120	.127	.133	.144	.147	.153	.160	.167	.180	.194	.207	.220	.244	.274
.042	.065	.072	.078	.084	.091	.098	.105	.111	.118	.125	.136	.138	.145	.152	.158	.172	.185	.198	.212	.235	.266
.035	.056	.064	.070	.076	.083	.090	.096	.103	.110	.116	.127	.130	.136	.143	.150	.163	.177	.190	.203	.227	.257
.032	.053	.060	.066	.072	.079	.086	.093	.099	.106	.113	.119	.126	.133	.139	.146	.160	.173	.186	.200	.223	.253
.028	.048	.055	.061	.068	.074	.081	.085	.094	.101	.108	.114	.121	.128	.135	.141	.155	.168	.182	.195	.218	.249
.025	.044	.052	.057	.064	.071	.077	.085	.091	.097	.104	.111	.118	.124	.131	.138	.151	.164	.178	.191	.214	.245
.022	.040	.048	.054	.060	.067	.074	.080	.087	.094	.100	.107	.114	.121	.127	.134	.147	.161	.174	.188	.211	.241
.020	.038	.045	.051	.058	.065	.071	.078	.085	.091	.098	.104	.111	.118	.125	.132	.145	.158	.172	.185	.208	.239
.018	.035	.043	.049	.055	.062	.069	.076	.082	.089	.096	.102	.109	.116	.122	.129	.143	.156	.169	.183	.206	.236

Table for finding the developed length of 90° bends. Subtract the correct figure in the table from the sum of the length of the legs. (Courtesy of The Glenn L. Martin Company)

2. When the outside dimensions of a bend are given, subtract one radius and one thickness for each dimension.

3. When there is a bend on each end of a dimension, the following procedure is followed. When the dimension is given outside, subtract

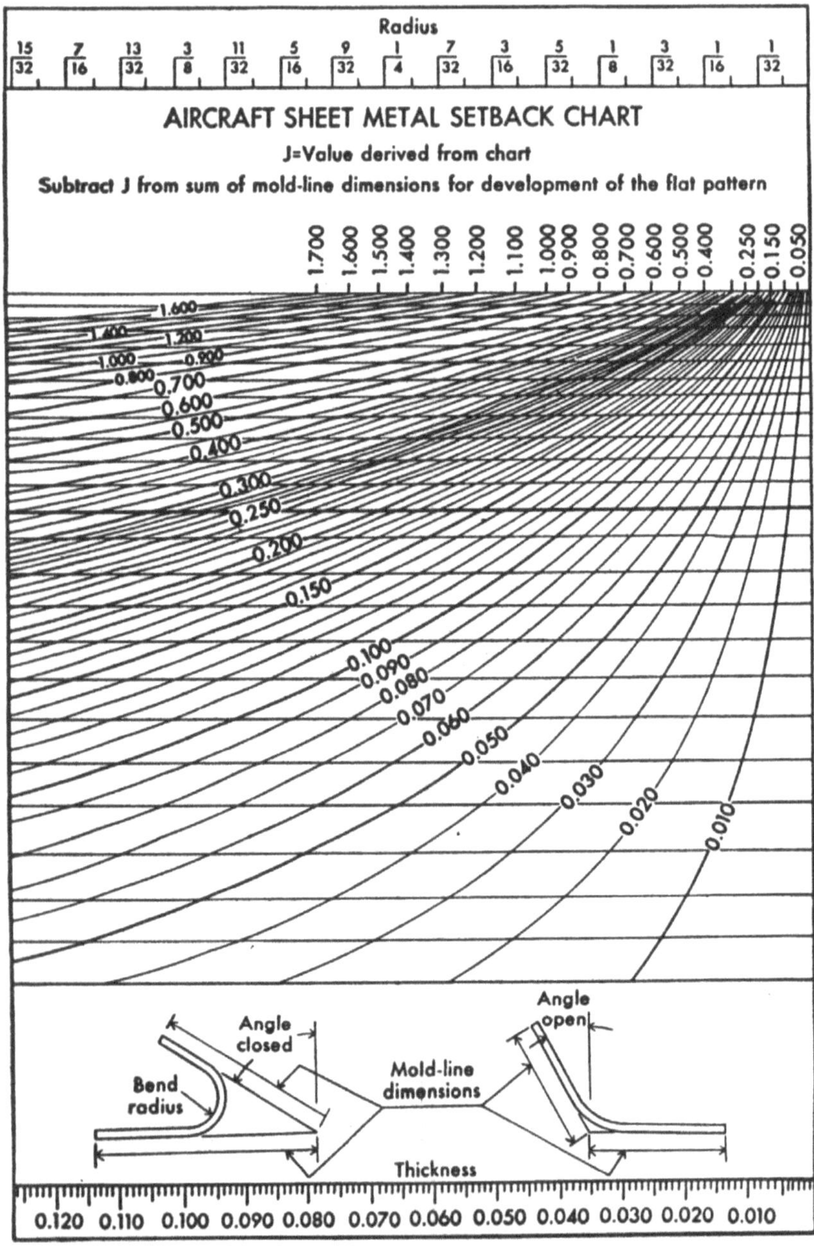

Fig. 105. Setback chart.

one radius and one thickness for each bend or a total of two radii and two thicknesses.

In the above when the inside dimensions are given, subtract only the two radii.

Setback charts have been designed, such as the one shown in Figure 105, to determine the setback without the use of the empirical formula illustrated in the preceding problem. In using the setback chart, a straightedge is laid across the chart from top to bottom, connecting the figures at the top representing the bend radius, with the proper figure at the bottom which shows the thickness of the material. The words "open" and "closed" will be noted along the right side of the chart. These words tell whether the angle is less than 90° or greater than 90°. If an angle is greater than 90°, it is called an open angle, while one less than 90° is called a closed angle. The rows of figures above and below 90° represent the number of degrees an angle is greater or less than 90°. Ninety degrees is represented by the heavy line near the center of the chart and parallel to the bottom of the chart. An angle of 80°, which is 10° less than 90° and therefore a closed angle, would be represented by the number, 10, above the 90° point. While an angle of 120°, which is 30° greater than 90° and therefore an open angle, would be represented by the figure, 30, below the 90° point.

If the bend radius is 3/8 in. and the thickness is 0.032, the straightedge should connect these points on the chart. For an angle of 60° the line opposite 30, which is above the 90° line since it is a closed angle, is located and followed across the chart to the point where it intersects the straightedge. This point falls between the line marked 0.600 and that marked 0.700 and is the required setback, or J value. Estimating this value since the point does not fall on a given line, the setback is found to be approximately 0.631. This amount is to be subtracted from the sum of the mold lines of the angle. The sum of the mold lines of the angle, as shown in Figure 106, is 3½, from which is subtracted 0.631, leaving 2.869, which is the developed length of the material necessary to form the flange. The setback chart is not as accurate as the empirical formula, but it is sufficiently accurate for most aircraft,

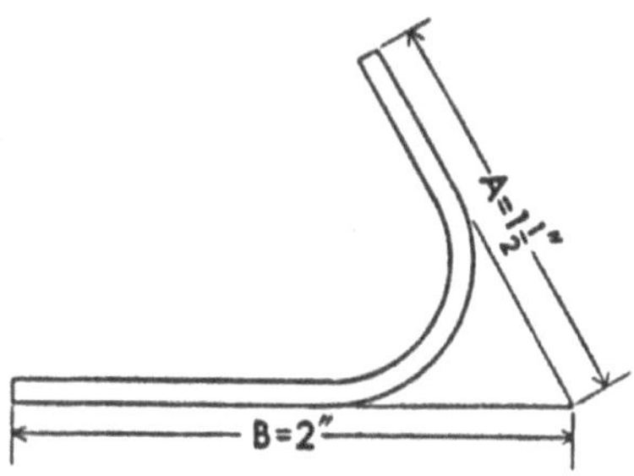

Fig. 106. An angle to show the use of setback.

sheet-metal work. The developed length will be within a few one-hundredths of an inch and usually falls within the tolerance limits.

Another type of chart, as shown in Figure 107, is also used to figure bend allowance. The first column, T, shows the thickness of the material.

COMBINED SETBACK AND BEND ALLOWANCE CHART

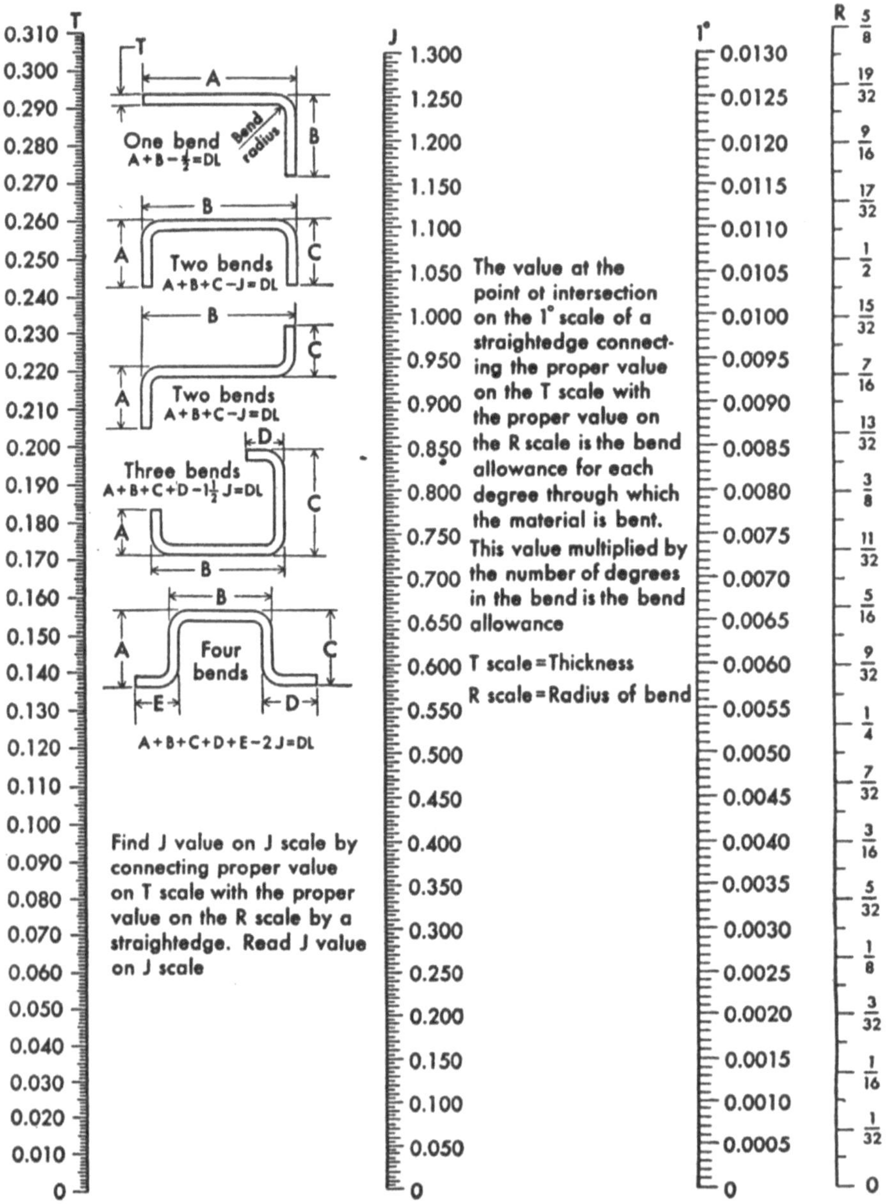

Fig. 107. Combination setback and bend allowance chart.

BEND ALLOWANCE

R = Radius of bend
T = Thickness
$x = A - C,\ y = B - C$
$C = (R + T)\cot\frac{D}{2}$
D = Angle between the flats
z = Bend allowance for one degree times the number of degrees in bend
$x + y + z$ = Developed Length (DL)

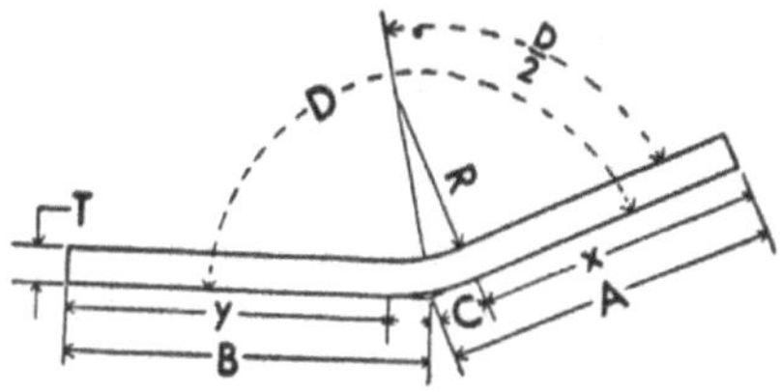

When angle D = 155°
Angle of bend = 180° − D
= 180° − 155° = 25°
R = 0.5625″, T = 0.031″
A = 1 3/16″, B = 1 1/4″
$\frac{D}{2} = 77°\ 30'$
$C = (R + T)\cot\frac{D}{2}$
= (0.5625 + 0.031)0.22169
= 0.1316
x = A − C
= 1.1875 − 0.1316
= 1.0559
y = B − C
= 1.25 − 0.1316
= 1.1184
z = 0.01005 (Table IV) × 25°
= 0.2512
Developed Length (DL) = x + y + z
= 1.0559 + 1.1184 + 0.2512
DL = 2.4255

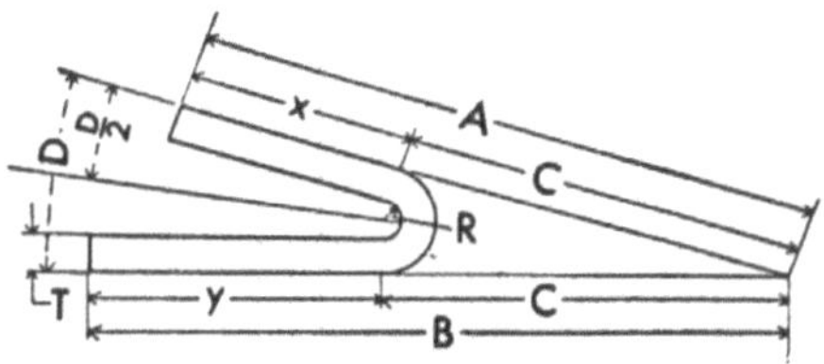

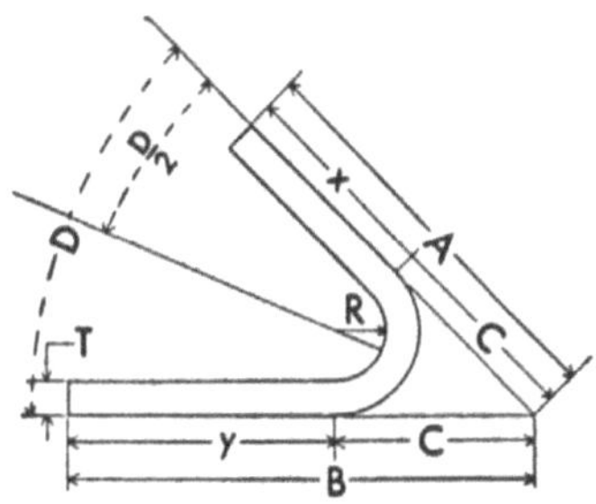

When angle D = 15°
Angle of bend = 180° − D
= 180° − 15° = 165°
R = 0.0625, T = 0.040
A = 2 1/4″, B = 2 1/2″
$\frac{D}{2} = 7°\ 30'$
$C = (R + T)\cot\frac{D}{2}$
= (0.0625 + 0.040)7.5957
= 0.7786
x = A − C
= 2.25 − 0.7786
= 1.4714
y = B − C
= 2.5 − 0.7786
= 1.7214
z = 0.0014 (from Table IV) × 165
= 0.2310
DL = x + y + z
= 1.4714 + 1.7214 + 0.2310
DL = 3.4238

When angle D = 45°
Angle of bend = 180° − D
= 180° − 45° = 135°
R = 0.1875, T = 0.064
A = 1 1/2″, B = 1 3/4″
$\frac{D}{2} = 22°\ 30'$
$C = (R + T)\cot\frac{D}{2}$
= (0.1825 + 0.064)2.4142
= 0.607
x = A − C
= 1.5 − 0.607
= 0.893
y = B − C
= 1.75 − 0.607
= 1.143
z = 0.00377 (from Table IV) × 135°
= 0.5089
DL = 0.893 + 1.143 + 0.5089
= 2.5449

Fig. 108. Finding the developed length (DL) by the use of the empirical formula and Table IV.

In the second column, J, is found the setback or J value for 90° bends. The values found on the chart are equal to twice the values to be used for a 90° bend. The third column, 1°, gives the value for the bend allowance for 1° for different combinations of bend radii and thickness. The fourth column, R, gives the bend radii.

To use this chart, a straightedge is placed across the chart from column T to column R, connecting the thickness of the material with the desired bend radius. Read the value in column J where the straightedge crosses the vertical line in the column to get a value which is double that for a 90° bend. For a single 90° bend, this value divided by 2 and subtracted from the sum of the sides of the angle, including the mold line, gives the approximate developed length of the flat pattern. Figure 107 A, B, C, D, and E illustrates the use of this part of the table. To find the bend allowance for each degree of bend, the straightedge is placed across the chart connecting the thickness with the desired bend radius, and the number is read from the vertical line in the column, 1°. This number is then multiplied by the number of degrees through which the metal is bent to find the amount of material necessary to make the bend. This number is added to the sums of the flats forming the sides of the angle to develop the flat pattern.

Figure 108 shows how to use Table IV in figuring the bend allowance from the empirical formula.

XV SHRINKING AND STRETCHING

The working of sheet metal by hand is almost an art. The workman must develop a sense of feel, which can only come by continued practice. It is necessary that the sheet-metal worker know exactly what to do and where to strike the metal to bring about the desired result. He must have a knowledge of the alloy with which he is working and its forming characteristics. A large part of the handwork done in aircraft factories consists of reworking parts which have been machine formed. This work may be stretching, shrinking, straightening, flattening, burring, or filing.

Cold-working on aluminum alloys brings about strain hardening. A part may become brittle due to extensive cold-working. Metals and alloys in the annealed condition will stand more working than when in the T condition. Most parts should be worked, if possible, in the annealed condition. If the handworking operation is to remove distortion caused by heat treating, the work should take place immediately after heat treating before age-hardening conditions have developed. After about one hour at 70° F., heat-treated material becomes difficult to work. For example, the alloy, 24S, when first heat-treated has a yield strength of approximately 22,000 lb. per sq. in., but within an hour this yield strength has increased to 28,000 lb. per sq. in. It will reach its maximum age-hardened condition in approximately one day after heat treatment. If it is necessary to delay forming, the material should be kept at or below 32° F. to retard age hardening. The alloys, 17S and 24S, may be kept in the soft condition for long periods of time at temperatures below 0° F.

The term, stretching, when used in connection with hand forming means that the material lengthens due to the operations performed upon it. Pure aluminum and alloys in the annealed condition are soft and malleable. The stretching action does not seem to weaken the material if it is not carried to excess. In stretching, the metal becomes thinner, just as a rubber band does when it is pulled out.

Fig. 109. Wrap-stretch forming an extruded aluminum alloy airplane part. (Courtesy The Glenn L. Martin Company)

Fig. 110. Forming aircraft skin on a stretching machine. (Courtesy The Glenn L. Martin Company)

Metal can be made to flow under the blows of a hammer. When the blows are applied in the proper manner, the metal can be made to flow or move in the desired direction. There are several ways to

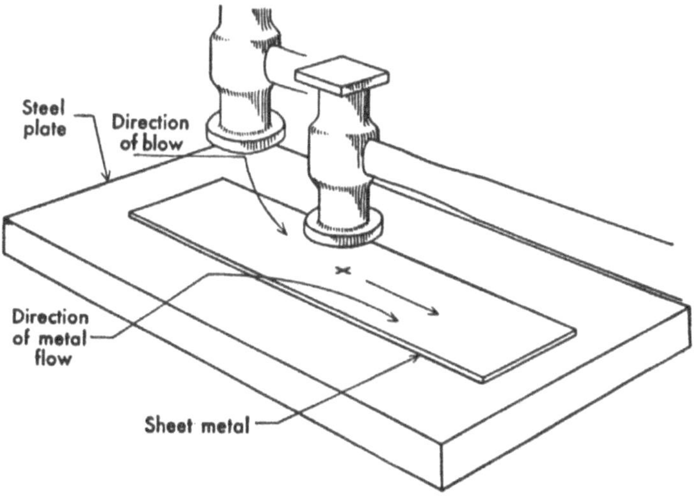

Fig. 111. When stretching metal by hammering on a metal plate the blows are struck with a circular motion.

stretch metal. The edges may be clamped and the sheet actually pulled as one would pull a rubber band, or it may be stretched by squeezing it between the hammer and the surface of an anvil or bench plate. When hammered, the metal immediately under the face of the hammer is

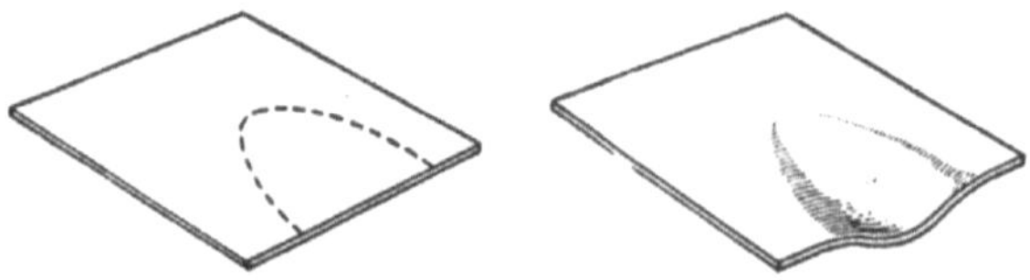

Fig. 112. Forming an air scoop.

forced outward. The face of the hammer must be properly shaped for the job, and the blows should be overlapping. The blows are struck with a circular motion in the direction it is desired that the metal move, as shown in Figure 111. Each blow of the hammer should be made

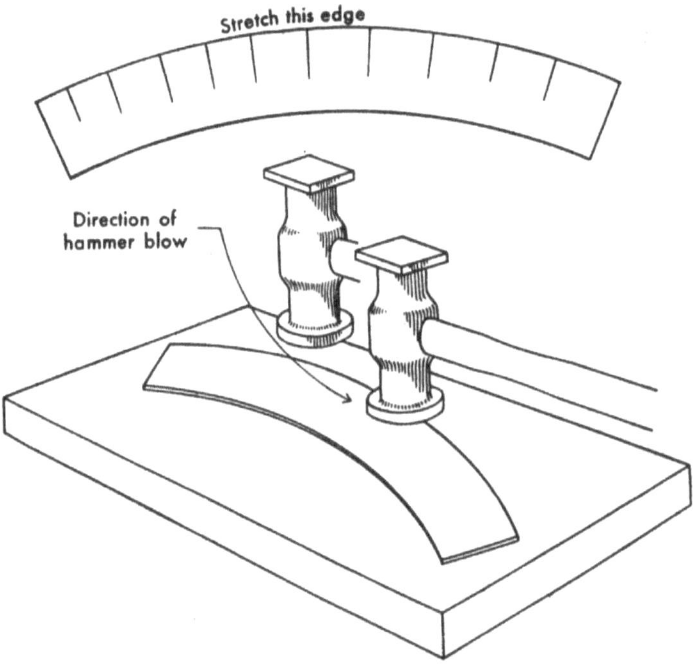

Fig. 113. Stretching one edge of a sheet to form a curve.

with a certain definite result in mind. The workman should know where he wants the metal to move and not simply hammer blindly.

An air scoop may be formed in the edge of a sheet by stretching the metal, as shown in Figure 112. A flat strip of metal may be lengthened by hammering it on a bench plate or other flat, polished surface, as shown in Figure 111. A straight strip of metal may be curved by stretching one edge, as shown in Figure 113. A piece of metal in the form of an angle, as shown in Figure 114, may be curved by stretching one side of the angle.

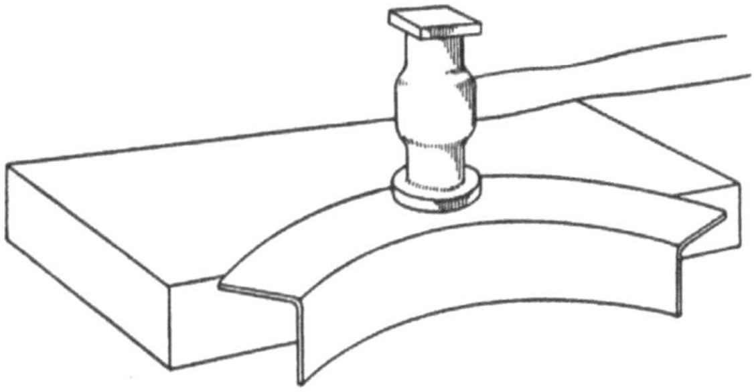

Fig. 114. Forming a curve on a flange by stretching one side.

It is easier to stretch metal than it is to shrink it. A V-block, as shown in Figure 115, should be available to the handworker in sheet metal. An angle may be curved by the use of a soft mallet and a V-block, as shown in Figure 116. One side of the angle is inserted in the slot in the V-block and hammered. The blows should be just heavy enough to

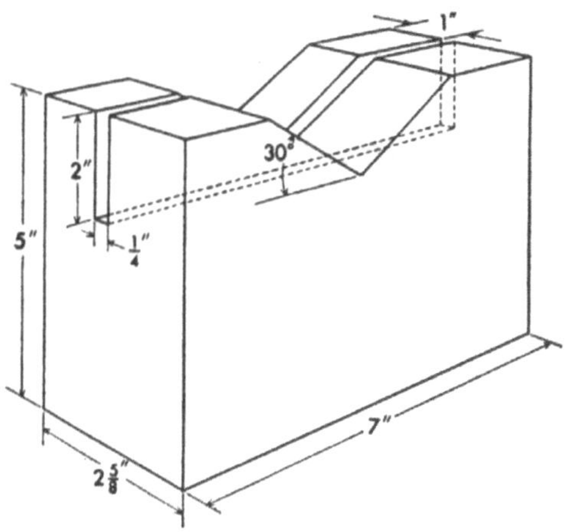

Fig. 115. A V-block.

obtain the desired results. Blows that are too heavy may cause buckling of the material or dents which are difficult to remove. Excessive hammering may so strain-harden the material that, if the hammering continues, cracks or breaks may result. The fewest possible blows should be used.

Hollows or dished sections may be formed in aluminum alloy sheets by working the sheet into a hollow with a soft mallet or hammer, as shown in Figure 117. The metal should not be clamped so tightly to the top of the forming block that it cannot flow into the depression.

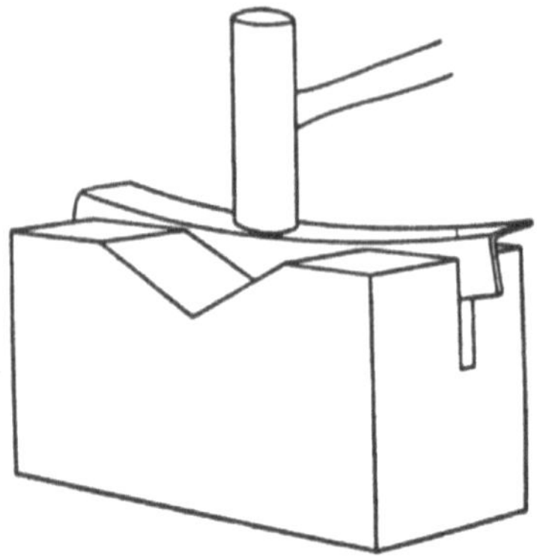

Fig. 116. Curving a flange by use of a V-block and mallet.

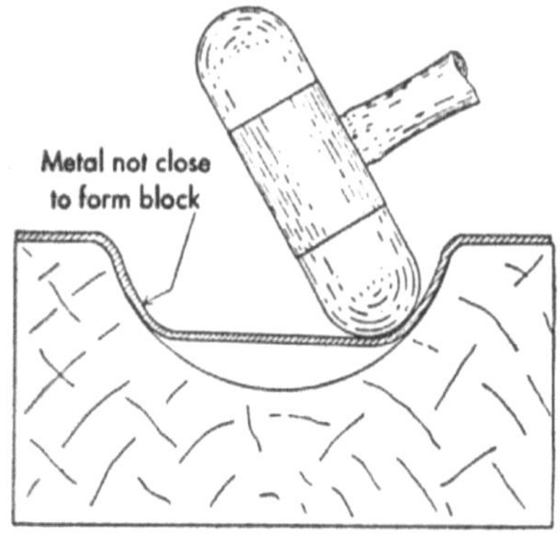

Fig. 117. Forming a hollow by use of a form block and mallet.

The deeper the hollow, the more the material will be stretched. The blow should be struck at the edge of the depression and not in the middle. The metal should be worked down the edge of the depression

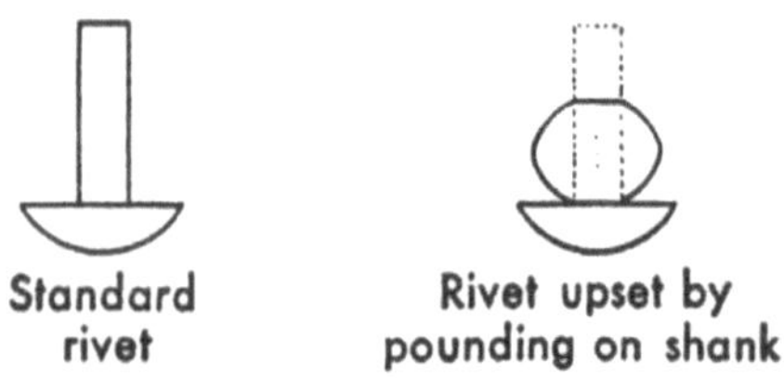

Fig. 118. A rivet before and after upsetting.

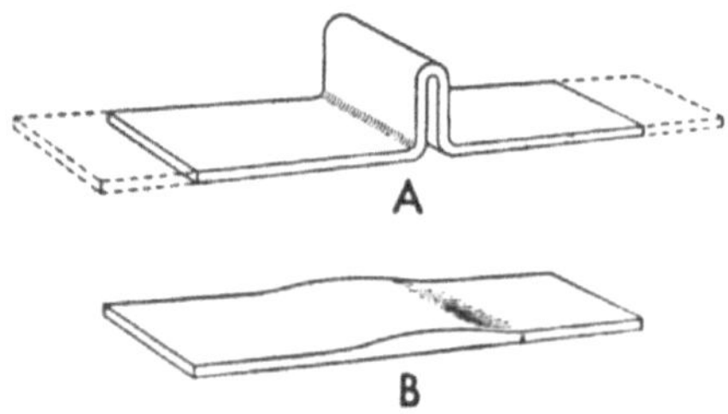

Fig. 119. The principle of shrinking. (A) A strip of metal bent into a fold. (B) The fold has been completely hammered out, leaving the sheet shorter and thicker.

to pull as much metal in from the top as possible so as to avoid thinning the sheet in the center of the depression.

Shrinking metal is just the opposite of stretching. The shrinking also causes hardening and sets up strains in the metal. Instead of the metal

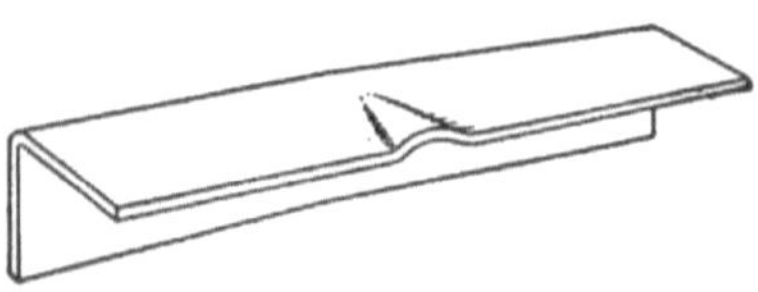

Fig. 120. A single crimp formed in the edge of a flange.

Fig. 121. Forming a curve in a flange by a series of crimps in one edge.

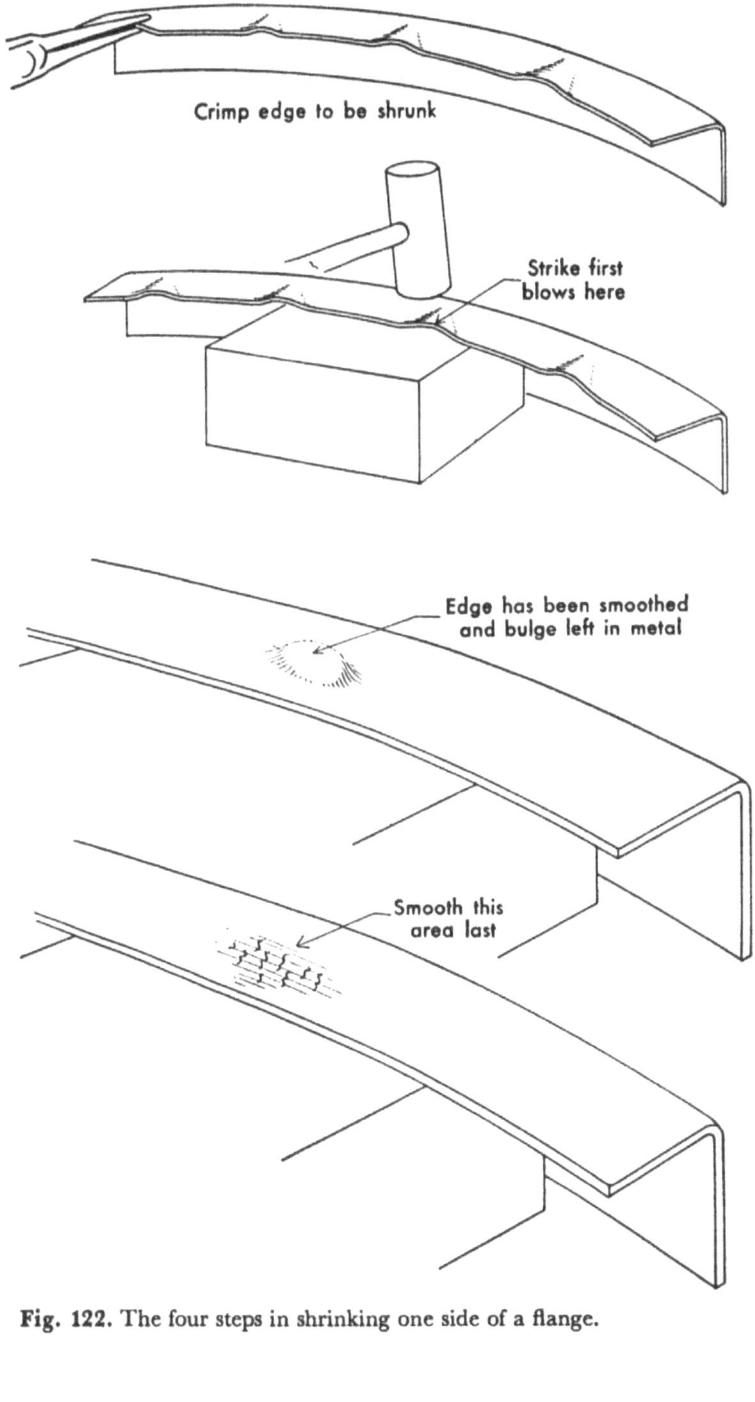

Fig. 122. The four steps in shrinking one side of a flange.

being thinned and lengthened as in stretching, shrinking causes the metal to become thicker and to decrease in area. The shrinking process is similar to the upsetting of a rivet. Figure 118 shows the rivet before and after upsetting. The rivet contains the same amount of material

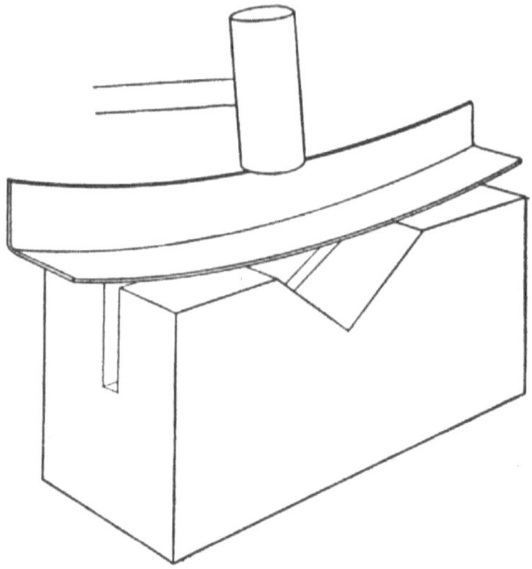

Fig. 123. A V-block used to shrink one side of a flange.

but is shorter and thicker than it was before the upsetting took place. If a strip of metal is bent as shown in Figure 119A and the bend hammered out by striking at the top of the fold, the metal in the sides of the fold is upset in a way similar to the upsetting of the metal in the

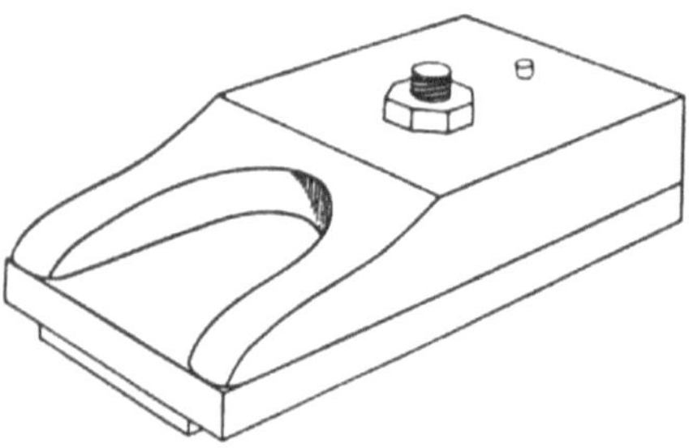

Fig. 124. A shrinking block.

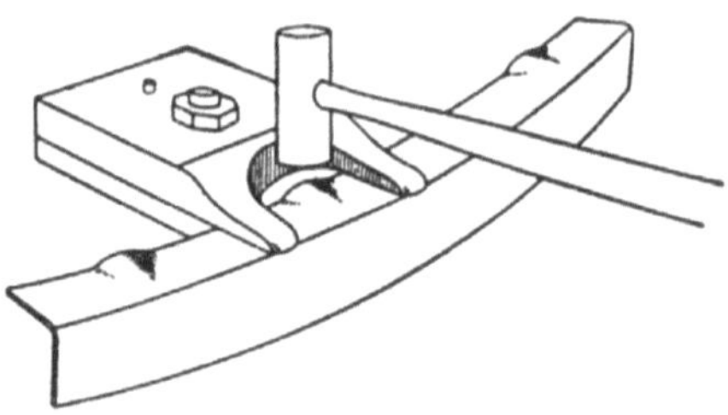

Fig. 125. Using a shrinking block to shrink one side of a flange.

rivet. As the fold is hammered completely out, as shown in Figure 119B, the strip becomes shortened, but is considerably thicker at the point of hammering. Severe bends of this type are not practical, but the illustration is used to show the principle of shrinking.

When actually forming a curve, a small crimp or bend is placed in the edge of a sheet, as shown in Figure 120, and then hammered out by striking on the top of the crimp. The resistance that the sheet offers on each side of the bend allows a part of the metal on each side of the fold to become upset, shrinking the edge of the sheet. A number of small folds or crimps may be placed in the edge of a flange to form a curve, as shown in Figure 121. These bends, when hammered out, will cause the edge of the flange to become shortened or shrunk. Figure 122 shows how this process is carried out to shrink one side of a flange.

The V-block may also be used to shrink one side of a flange, as shown in Figure 123. This process is the opposite of that used to stretch one side of the flange. When shrinking one side of a flange in this manner, the flange will have a tendency to close. When the flange is opened, the tendency is to straighten the curve which has been formed in the side of the flange. The curved part should be clamped between two blocks while the flange is opened.

A shrinking block made of steel or other suitable material, such as shown in Figure 124, is useful in shrinking flanged parts. The crimps placed in the edge of the part to be shrunk are clamped in the shrinking block and hammered out, as shown in Figure 125. Flanges may be made by the use of a proper form block and backing block, as shown in Figure 126. The blocks with the material to be formed between them are clamped in a vise. Using the correct mallet, the material is bent into

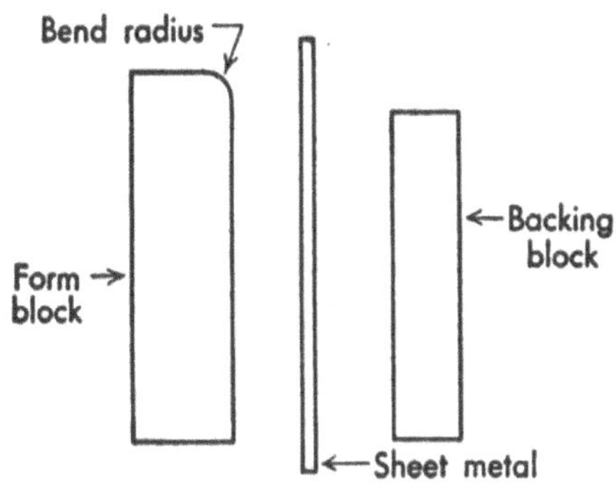

Fig. 126. A form block, the material to be formed, and a backing block.

place over the form block. The form block may be made from hardwood, Masonite, Micarta, plastic, or suitable metal. The form block should have the proper radius on the edge over which the metal is to be formed, and a sufficient allowance should be made for spring back. The metal should be formed over the block by use of a cross-peen wooden mallet, and the blows struck close to the form block, as shown in Figure 127. Curved flanges which are difficult to form may be made in this manner. In forming a curved flange the metal must be shrunk. The excess metal should be supported by a backing-up or shrinking tool to prevent sharp wrinkles from forming, as shown in Figure 128. This tool also assists in holding the metal tightly to the form block.

Problem 1. Obtain a strip of 2S aluminum approximately 1 in. wide and 4 in. long having a thickness of 0.125 in. Place the piece of metal

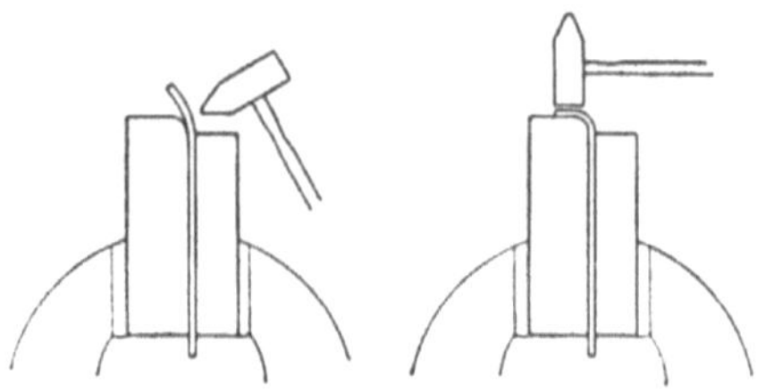

Fig. 127. Forming sheet metal by use of a form block, backing block, and wooden mallet.

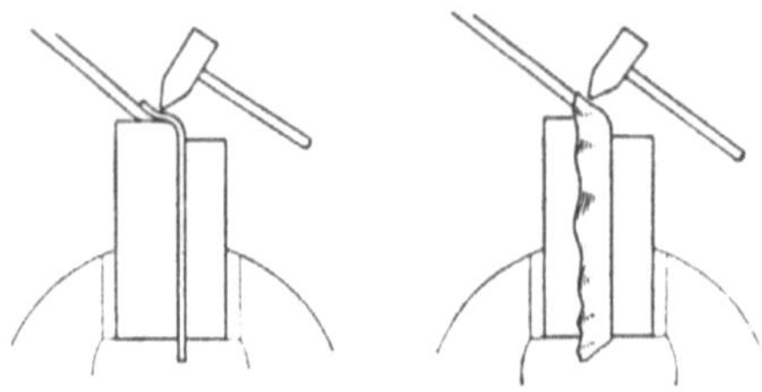

Fig. 128. Using a backing-up tool to prevent wrinkles when forming a curved flange.

on the smooth surface of an anvil or bench plate and, using a finishing or planishing hammer, stretch the metal in length. The face of the hammer should be polished and flat. The hammer should meet the metal squarely, but the blows should be struck with a circular motion so that the hammer forces the metal to move in the proper direction. Begin at the middle of the piece and work toward each end. Try to stretch the piece, increasing its length by 20 per cent. Do not allow the piece to increase in width. If the metal has a tendency to increase in width, change the direction of the blows so that they strike the metal diagonally toward the center, working the metal in from the edge. After the metal has increased 20 per cent in length, it should be of uniform thickness from end to end with practically no increase in width. Continue hammering the metal until cracks and breaks begin to appear as a result of work-hardening. The increase in hardness of the metal can usually be determined by the feel of the metal under the hammer.

Problem 2. Obtain a piece of 2S aluminum approximately 1 in. wide and 4 in. long, having a thickness of 0.063 in. By using the same tools as in Problem 1, try to increase the width of this strip 20 per cent without increasing its length. When finished, the thickness should be uniform throughout. Continue hammering the piece until cracks begin to appear due to strain hardening.

Problem 3. Repeat the exercises given in Problems 1 and 2 using 17S–O and 24S–O aluminum sheet. Note the difference in the feel of the metals under the hammer and the difference between the stretching qualities of these alloys and pure aluminum.

Problem 4. Repeat the exercises in Problems 1 and 2 using strips of 17S–T and 24S–T aluminum sheet, having a thickness of 0.063 in. Note the increased difficulty in forming. The heat-treated alloys are

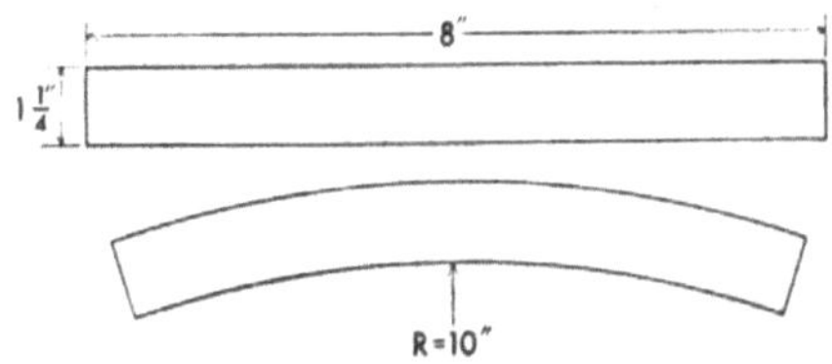

Fig. 129. Forming a curve by stretching one side.

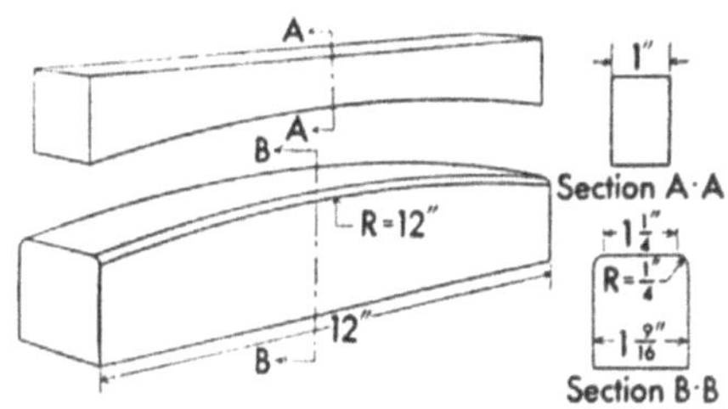

Fig. 130. Form and backing blocks used to form a curved channel.

much harder and the metal is more elastic, which tends to cause it to return to its original shape after each hammer blow.

Problem 5. Using a flat strip of 2S sheet 1¼ in. wide and 8 in. long having a thickness of about 0.063 in., form the sheet into a curve, as shown in Figure 129, by stretching one edge. Do not try to stretch just the edge. The stretching should be graduated from one edge to the other. Try to form this curve without any increase or decrease in the width of the strip. Try this exercise also with strips of 17S–O and 24S–T.

Problem 6. Prepare a form and backing block, as shown in Figure 130, and form a curved channel from 2S aluminum having a thickness of 0.040 in. The finished angle should conform to the dimensions given in Figure 131 without trimming after forming. This means that bend allowance will have to be figured.

Problem 7. Form a small air scoop in the edge of a piece of 24S–O aluminum having a thickness of 0.040 in. It will be necessary to prepare a suitable form block and stretch the material to conform with the depression in the block. The clamps should not be tight enough to prevent the metal from flowing into the scoop during the forming operation.

Problem 8. Form a right angle in a piece of 2S aluminum having the dimensions shown in Figure 133, making the proper allowance for the bend before forming. The proper form and backing block should be prepared. Form the flange over the form block. To form the curve,

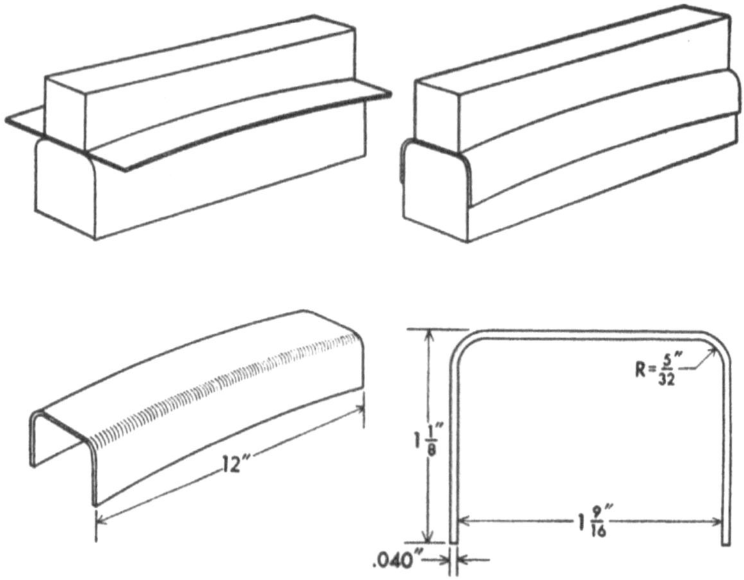

Fig. 131. Steps in forming a curved channel.

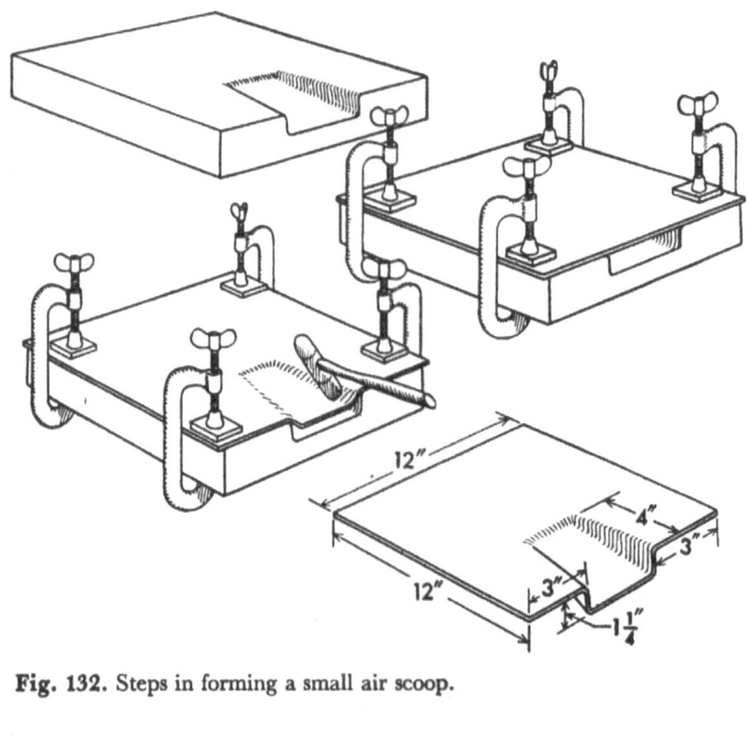

Fig. 132. Steps in forming a small air scoop.

remove from the form block and stretch one flange by hammering on an anvil or a bench plate. Continue the stretching until signs of failure appear in the metal. Try this exercise with 24S–O aluminum sheet.

Problem 9. Using a piece of 2S aluminum sheet having a thickness of 0.040 in., form a curved flange, as shown in Figure 134. First form a flange similar to the one in Problem 8. Then form the flange into a curve by shrinking one side of the angle. Crimp the edge and hammer out the crimps while the metal is clamped in a shrinking block. Prepare a similar flange and form into a curve by shrinking one side of the flange on a V-block, as shown in Figure 123.

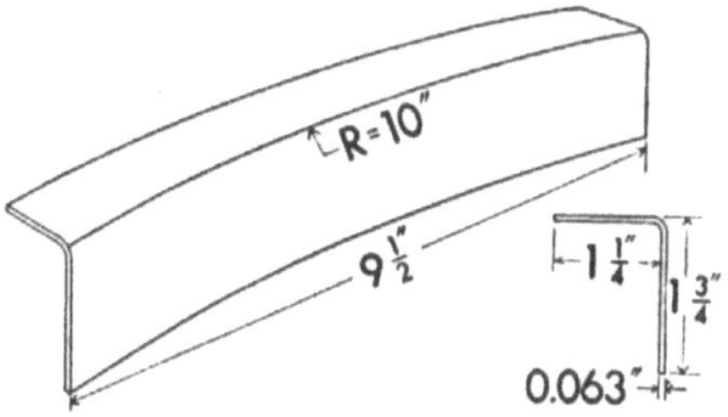

Fig. 133. A curved flange formed by stretching one edge.

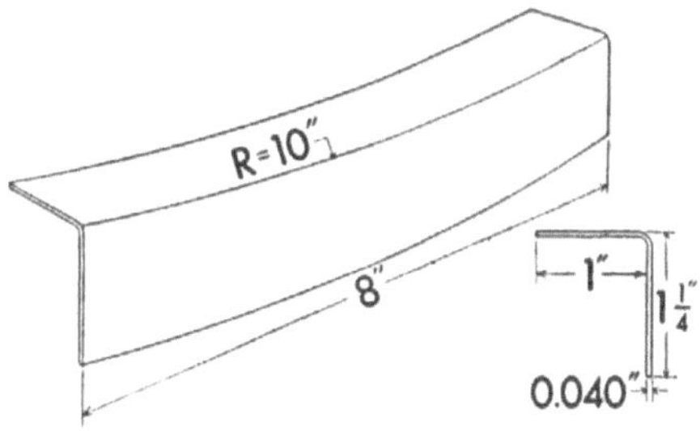

Fig. 134. A curved flange formed by shrinking one edge.

Problem 10. Using a strip of 2S aluminum ½ in. wide and 12 in. long having a thickness of 0.125 in., form a curve, as shown in Figure 135. Any wrinkles that form should be hammered out on a bench plate.

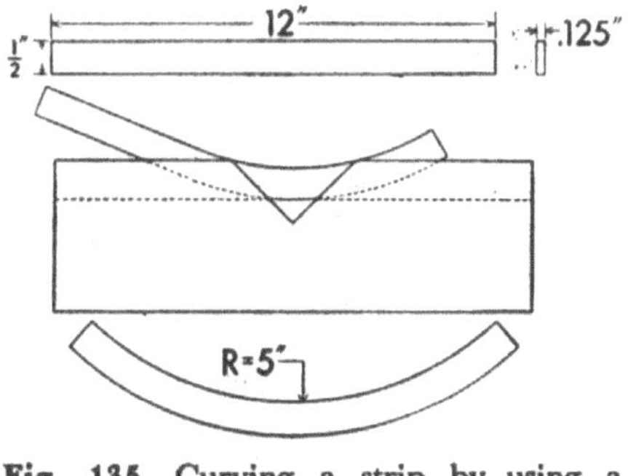

Fig. 135. Curving a strip by using a V-block and mallet.

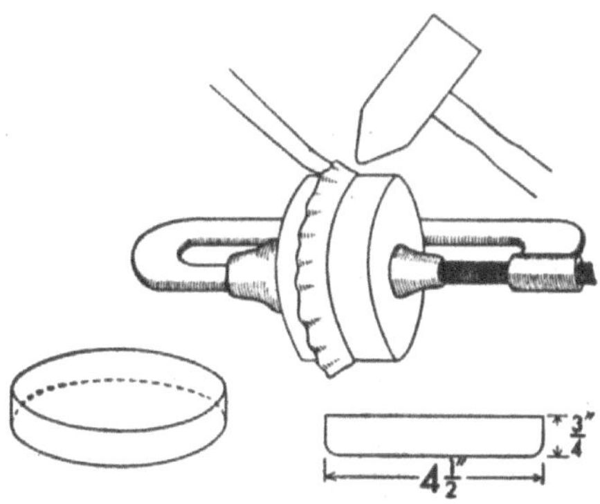

Fig. 136. Forming a shallow dish. Note the use of the backing-up tool.

Problem 11. By the use of the proper blocks, form a shallow dish, as shown in Figure 136, from 2S aluminum having a thickness of 0.040 in. Gradually shape the material completely around the block, taking care to prevent wrinkles, particularly in the flange. Use a backing-up tool.

Problem 12. Using a piece of 24S–O aluminum having a thickness of 0.040 in. form a double curved flange as shown in Figure 137. Note that one part of the flange is stretched and the other is shrunk.

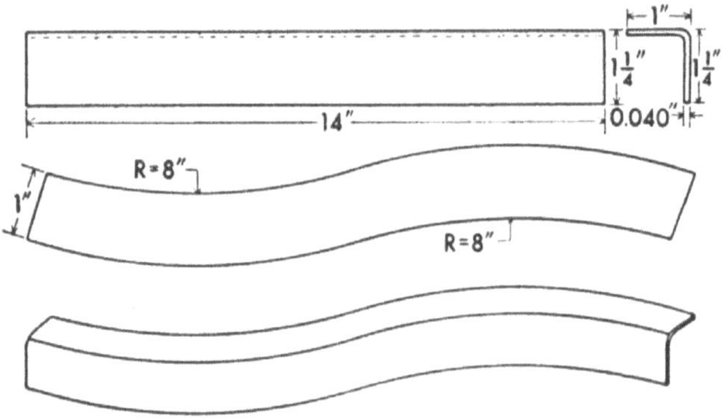

Fig. 137. Forming a double curved flange.

The following series of photographs[1] shows the methods used in hand forming a piece of aluminum alloy having a number of different curves. The forming of this piece includes almost all of the operations encountered in hand forming.

[1] Figures 138 through 154 are reproduced by the courtesy of The Glenn L. Martin Company.

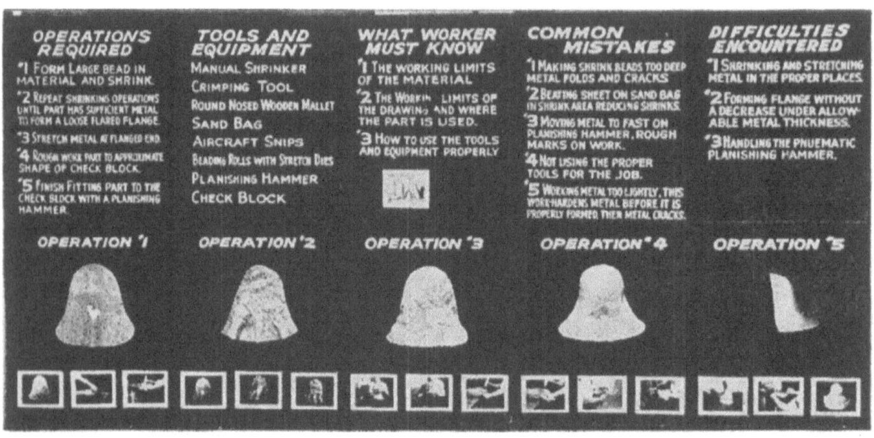

Fig. 138. The setup used in one of the large aircraft factories to show the various steps in forming sheet metal by hand. Figs. 139 to 154 show the detailed steps in forming this part.

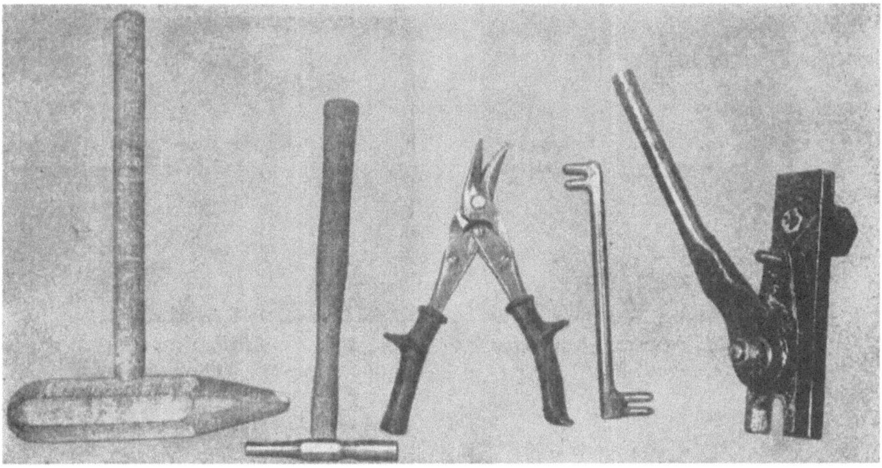

Fig. 139. Hand tools required for the job: a round-nose wooden mallet; planishing or finishing hammer; pair of aircraft snips; a crimping tool; and a shrinking block with wrench for tightening. The proper form and check block and sandbag should be at hand.

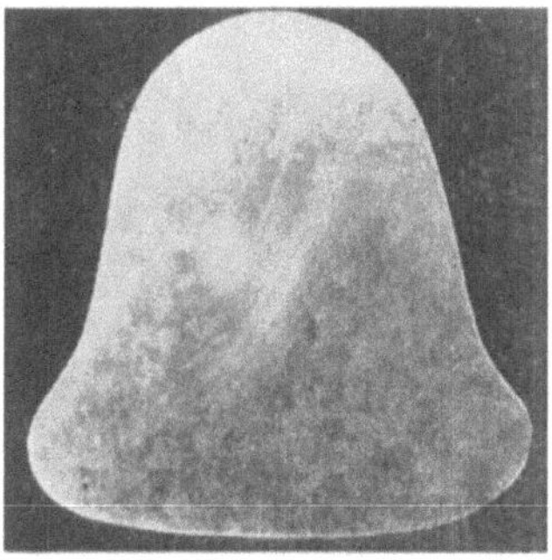

Fig. 140. The flat pattern which was cut from 52S–O aluminum alloy.

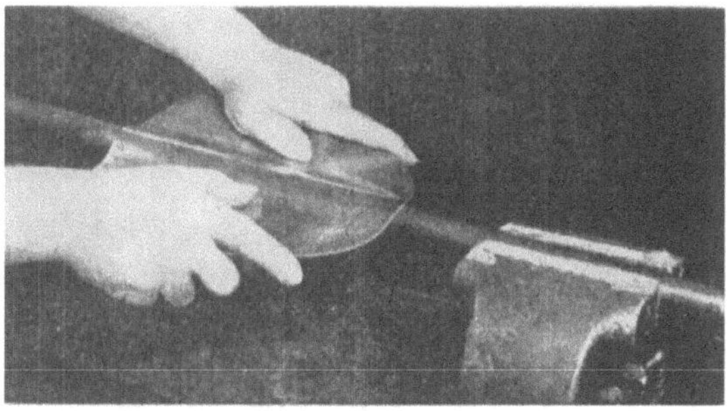

Fig. 141. Forming a crimp in the material, using a round bar clamped in a vise. The crimp is formed by hand.

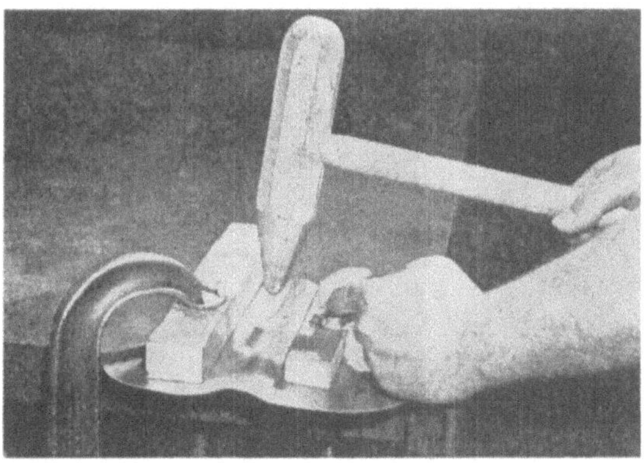

Fig. 142. The metal is clamped to a bench and the pointed end of a mallet is used to hammer out part of the crimp, shrinking the metal at the top of the sheet.

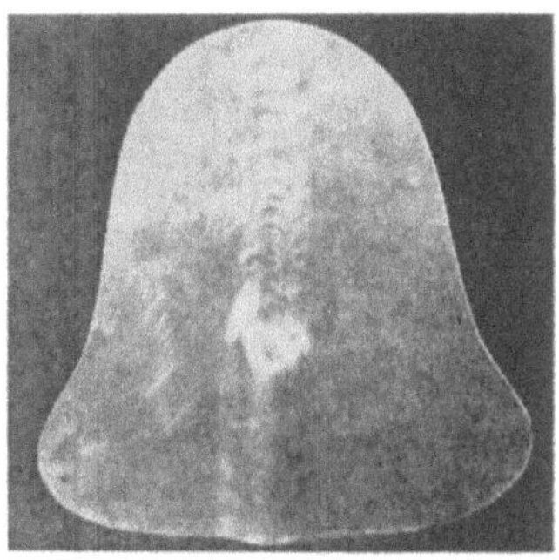

Fig. 143. Part of the crimp has been removed, shrinking the narrow part of the sheet.

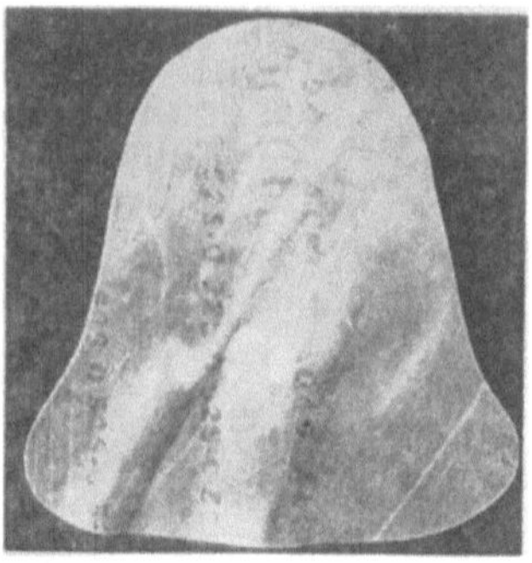

Fig. 144. A second crimp partially removed.

Fig. 145. A third crimp partially removed. The narrow part of the sheet is shrunk to a considerable extent.

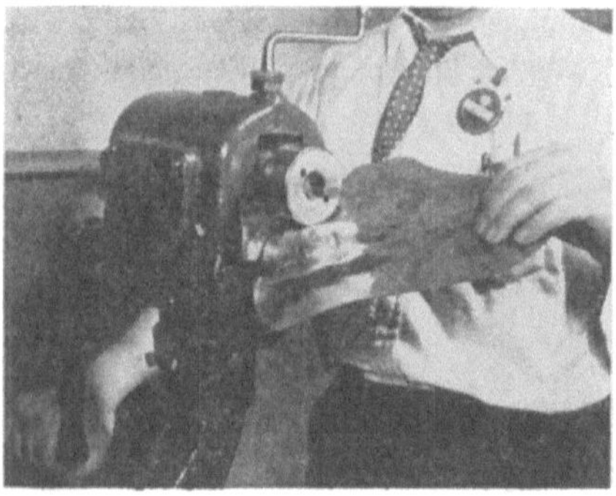

Fig. 146. The bottom of the sheet is stretched on beading rolls with stretcher dies. If beading rolls are not available, this stretching may be done by the use of a sandbag and mallet.

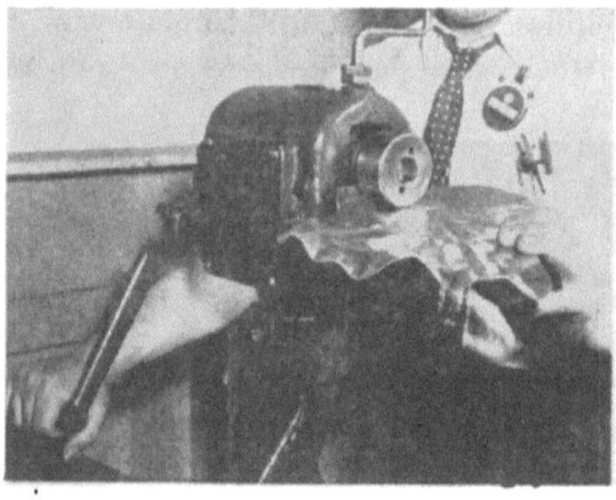

Fig. 147. The bottom of the sheet stretched, forming a wavy edge.

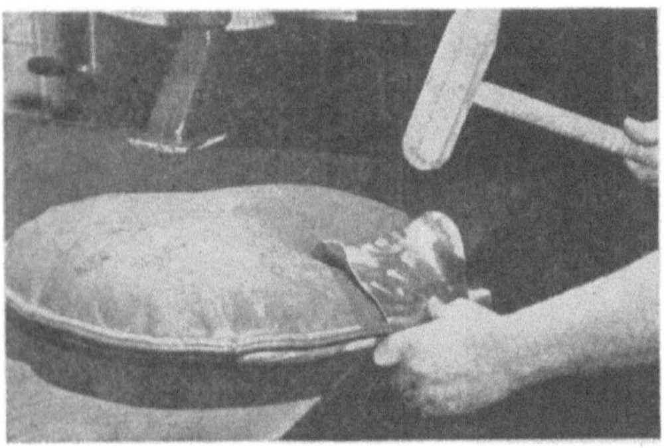

Fig. 148. A sandbag is used to pound out the wrinkles formed by stretching the bottom of the sheet. Some additional stretching is done in this operation.

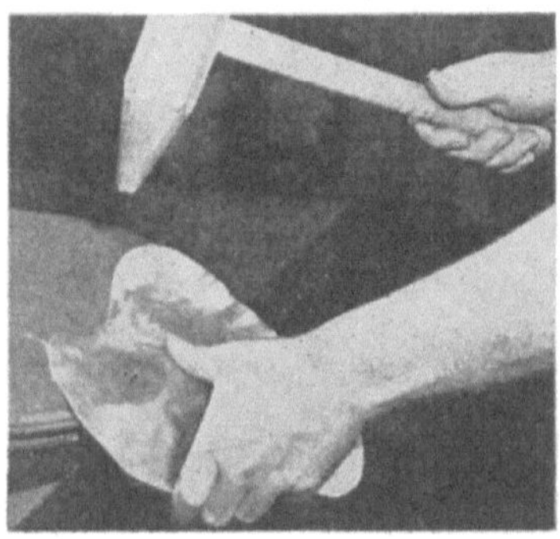

Fig. 149. The small end of the mallet is used to further stretch the flange.

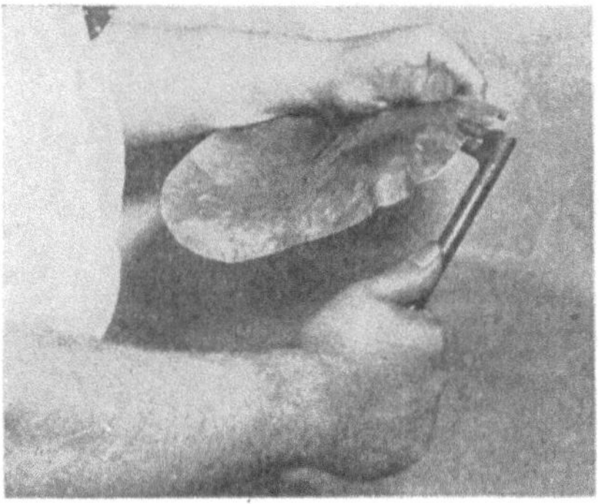

Fig. 150. A hand-crimping tool is used to crimp the top of the sheet for shrinking. This operation may be performed with round-nose pliers.

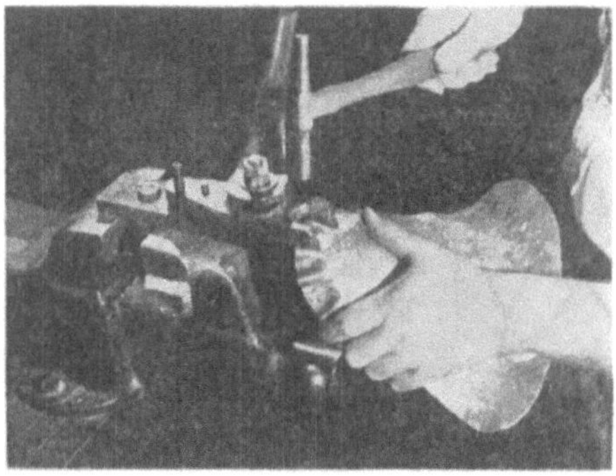

Fig. 151. The shrinking block is clamped in a vise. The material at the top of the sheet is shrunk by clamping in the shrinking block and hammering out with a planishing hammer.

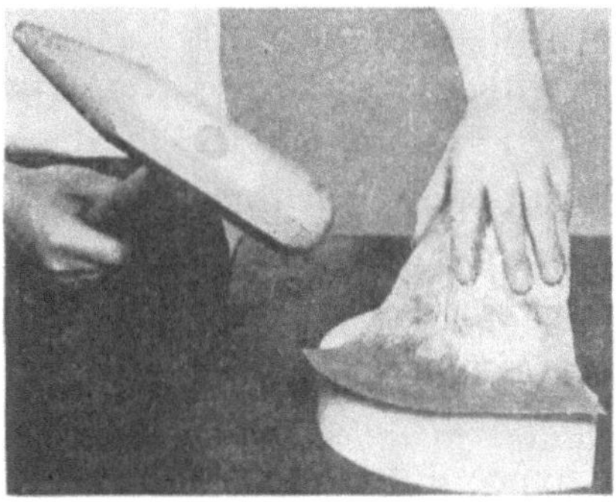

Fig. 152. The sheet is being formed over the form block. Care must be taken to strike all blows with the proper force and in the proper place.

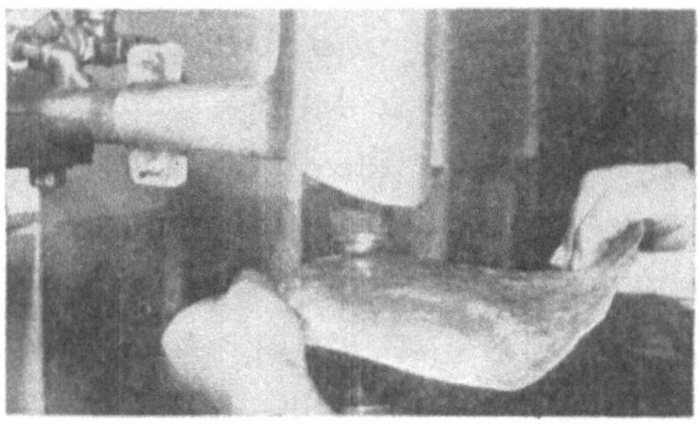

Fig. 153. The sheet is smoothed by the use of a pneumatic planishing hammer. The Yoder hammer is also commonly used in this operation. This may also be done by hand with the proper planishing hammer and a form block.

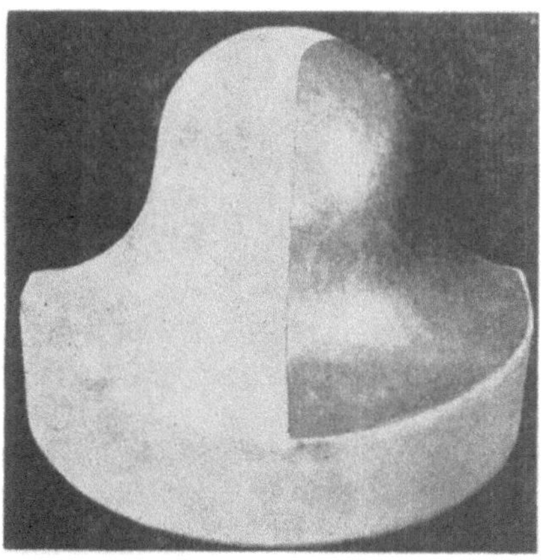

Fig. 154. The part trimmed and fitted to the check block.

XVI FORMING ALUMINUM AND ALUMINUM ALLOYS

Aluminum is one of the most workable of the metals used in aircraft fabrication. The workability of a metal is that property which allows it to be bent, formed, drawn, or spun without developing cracks or becoming weakened. Unlike the ferrous metals, aluminum and its alloys do not appear to have a definite yield point. However, there is a point beyond which the force necessary to deform the aluminum is not in direct proportion to the amount of deformation. The sheet-metal worker should not think of aluminum as being a simple material, but should realize that there are a large number of aluminum alloys having different forming characteristics due to their difference in hardness, springiness,

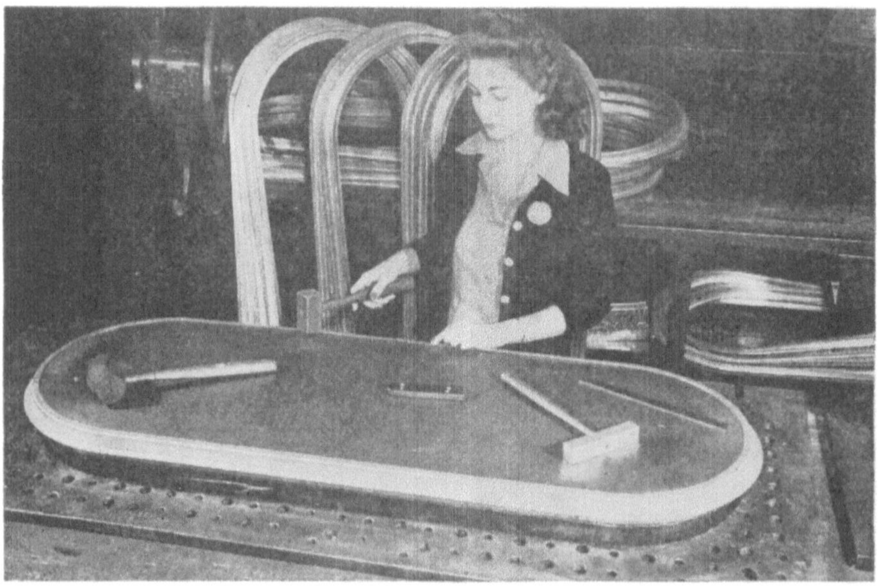

Fig. 155. Hand-finishing machine-formed parts that have been distorted by heat treatment. Note distorted parts in background. (Courtesy The Glenn L. Martin Company)

and ductility. The unit force used in forming aluminum has been defined as that force which produces a permanent set of 2 per cent of the original dimensions. This force is defined as the yield force or yield strength of the material.

Commercially pure aluminum, known as 2S, is noted for the ease with which it may be drawn or formed. This material in the fully

Fig. 156. Hand-finishing a nose piece using a wood tool and mallet. (Courtesy The Glenn L. Martin Company)

annealed condition may be subjected to a series of forming or drawing operations without being re-annealed. 3S is slightly more difficult to work, but all of the aluminum alloys may be worked at least to some extent in any of their various tempers. None of the alloys works as easily as pure aluminum.

Any operation which changes the shape of a piece of metal is known as forming. Forming includes bending, drawing, stretching, shrinking, spinning, and forging. Deep-drawing operations are usually performed with a punch and die. The material does not stretch evenly but flows at different rates, depending upon the shape of the object being formed. This may be seen in Figure 158. The lines shown on the formed parts

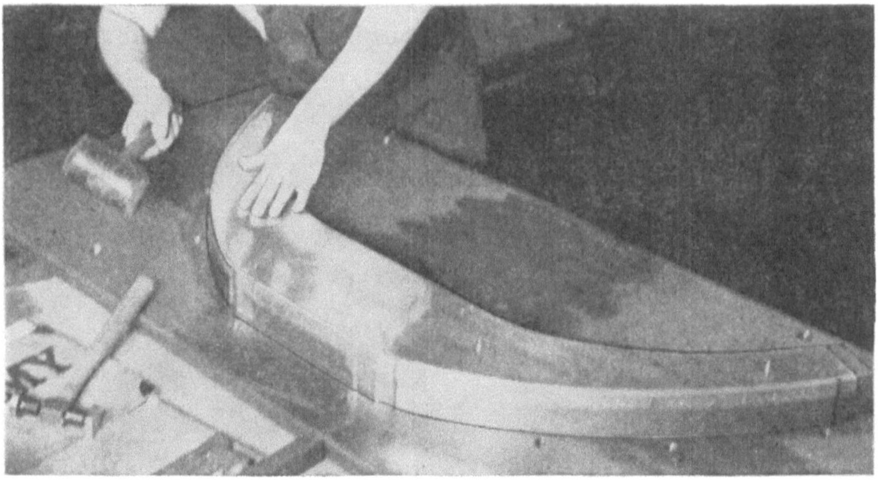

Fig. 157. Hand-finishing a machine-formed part. (Courtesy The Glenn L. Martin Company)

were drawn parallel and at equal distances apart on the flat blank. The lines on the finished part show the direction and distance in which the metal flowed. In stretching aluminum sheet to form a part such as an engine cowling, the strain lines may be seen as the metal is stretched.

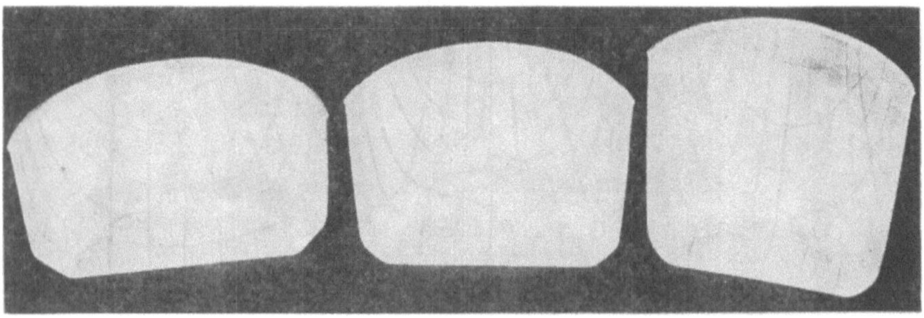

Fig. 158. Formed aluminum. The lines indicate the flow of the metal. (Courtesy Aluminum Company of America)

When aluminum is rolled, a grain structure is set up similar to the grain in a piece of wood. Like a piece of wood, the aluminum sheet is stronger at the bend line when the bend takes place across this grain than when it takes place parallel to it. The sheet-metal worker should

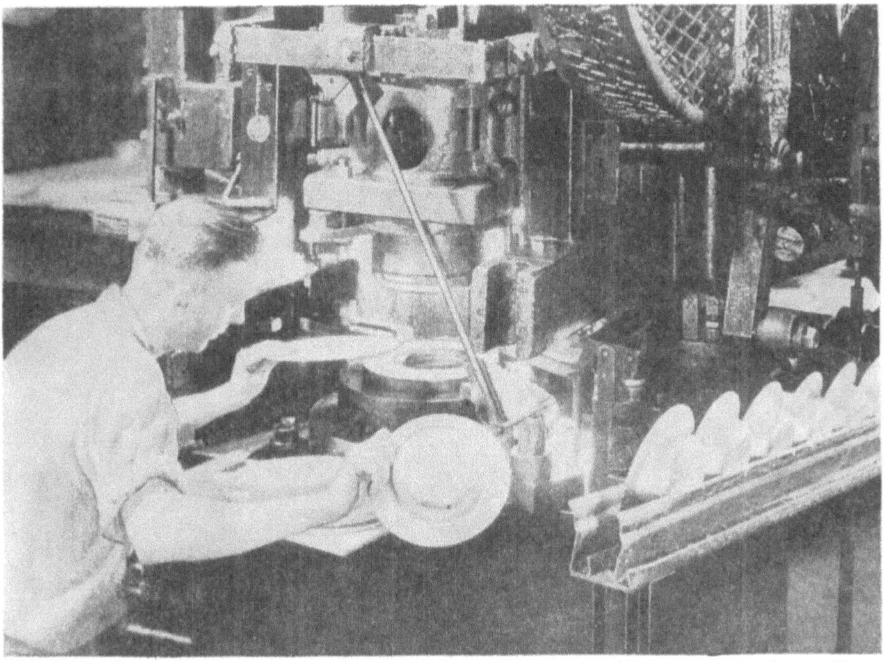

Fig. 159. The first drawing operation on a deep-drawn part. (Courtesy Aluminum Company of America)

Fig. 160. A sheet of aluminum alloy placed over forming tools. (Courtesy Aluminum Company of America)

Fig. 161. The same sheet as in Fig. 160 after removal from the hydraulic press. (Courtesy Aluminum Company of America)

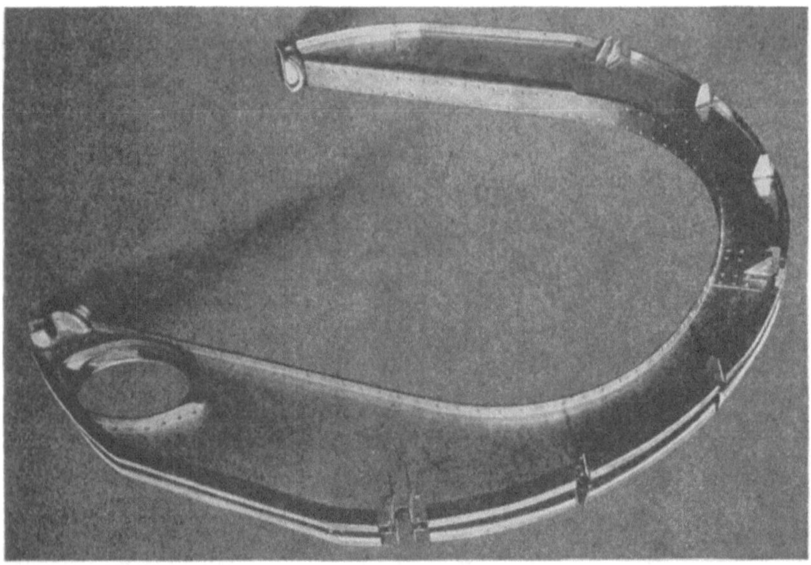

Fig. 162. The parts of this bulkhead were formed on a hydraulic press. (Courtesy Aluminum Company of America)

keep this in mind when working aluminum or aluminum alloy sheet. The grain lines run in the direction in which rolling takes place. Figure 159 shows the first drawing operation performed on an aluminum blank in forming a deep-drawn part. These parts are subjected to successive drawing operations, each operation deepening the part. Severe

Fig. 163. Spinning an aluminum part. (Courtesy Aluminum Company of America)

drawing is usually carried out in a series of operations. Figure 160 shows a sheet of aluminum alloy placed over forming tools or dies before placing in a hydraulic press. Figure 161 shows the same sheet after removal from the press. The press has forced the rubber down over the dies. The sharp edge of the die actually cuts the part from the sheet.

Such parts as the bulkhead shown in Figure 162 may be formed on a hydraulic press. The parts formed need only to be drilled and riveted in place to complete the part.

Spinning operations show plainly the workability of aluminum. Figure 163 shows the spinning of an aluminum part by use of a wooden spinning tool. Figure 164 shows steel tools used for spinning.

Aluminum alloys may also be forged, which means hammering into the desired shape. Hand-forming is closely related to forging. Heavy forging is done by machinery; light forging may be done with a heavy hammer, placing the part on an anvil and using properly shaped tools.

Fig. 164. Steel tools used for spinning aluminum. (Courtesy Aluminum Company of America)

Fig. 165. Hand-forming an extruded part. (Courtesy The Glenn L. Martin Company)

When forming light sheet or extruded parts by hand, the operation is called hand-forming rather than forging. The hand-forming operation shown in Figure 165 is similar to forging. The extruded part has been bent by hand around the form block and clamped into place. The sheet-metal worker is forming the metal close to the form block by the use of

Fig. 166. Forged parts formed by a drop hammer. (Courtesy Aluminum Company of America)

a wooden block and rubber mallet. Figure 166 shows forged parts formed by a drop hammer.

The aircraft sheet-metal worker should keep in mind the forming characteristics of the material with which he is working. He should be sure that proper bend allowance has been made, that the material has not been stretched far enough to cause excessive thinning, and that the material has not become work hardened during the forming process.

Problem 1. To form a 90° angle, obtain a piece of 2S aluminum sheet approximately 6 in. long and 3 in. wide, having a thickness of 0.040 in. Using the proper form and backing blocks, make a 1-in. flange along the long side of the sheet as follows. Determine the bend line and clamp the sheet between the form and backing blocks shown in Figure 167. Using a mallet made of soft wood shape the metal over the form block. Do not try to form the angle completely at any point. Work the metal gradually and evenly into place. The first blows should be struck as near the block

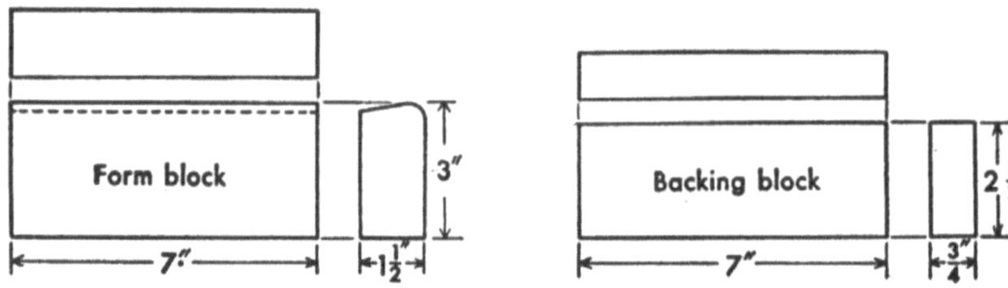

Fig. 167. A form block and backing block used to form a flange.

as possible. Do not bend the outer part of the flange to the form block ahead of the metal closer to the bend. If wrinkles tend to form, they should be removed by the use of a backing-up tool. A cross-peen wooden mallet may be used to start the bends, and the flat face of the mallet used to finish them. Check the angle with a trysquare to see that it is even throughout. The flange should be exactly 1 in. in width. Do not

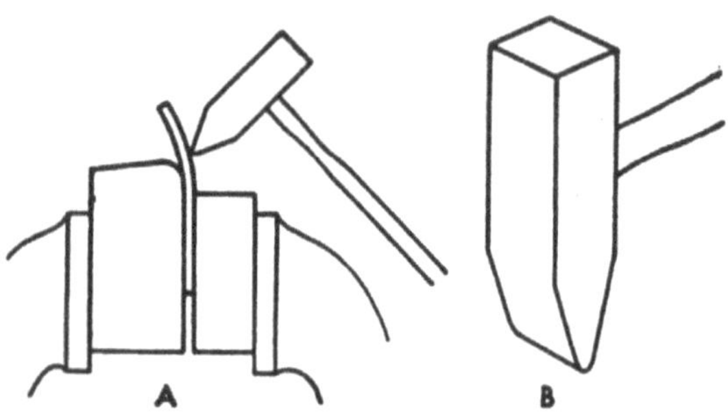

Fig. 168. The sheet is clamped in place, and forming started by use of a cross-peen wooden mallet.

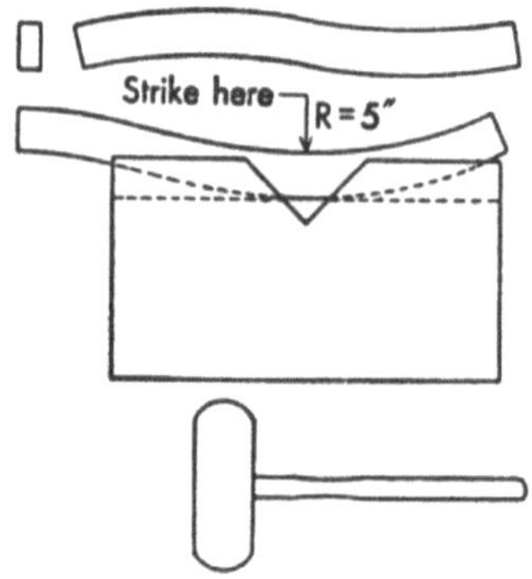

Fig. 169. Forming a double curve by use of a V-block.

strike glancing blows, as this will stretch the metal over the edge of the block and may thin it at that point.

Problem 2. Using 24S–O aluminum alloy sheet having a thickness of 0.040 in., repeat Problem 1. Notice the difference in reaction between this alloy and the 2S aluminum.

Problem 3. Repeat Problem 1, using 24S–T sheet having a thickness of 0.040 in. Note the difference in reaction of this alloy and the 2S and

24S–O. This material is much harder, and there is more tendency to spring back. Difficulty may be encountered in making the material lie close to the form block. It may be necessary to bevel the form block slightly, as shown in Figure 167, so that the angle may be bent over to allow for springback. It will be necessary to strike much heavier blows

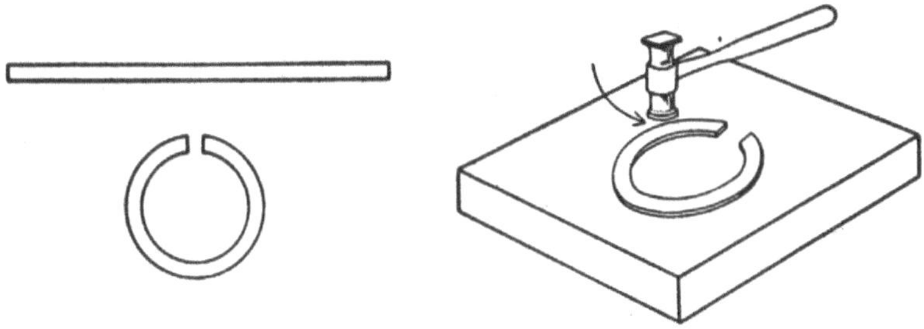

Fig. 170. Forming a circle from a strip of aluminum alloy.

in forming this material. The alloy, 24S, is usually formed in the annealed condition.

Problem 4. In 2S sheet approximately 10 in. long and ½ in. wide, having a thickness of 0.063 in., form a double curve as shown in Figure 169. Each curve should have an inside radius of approximately

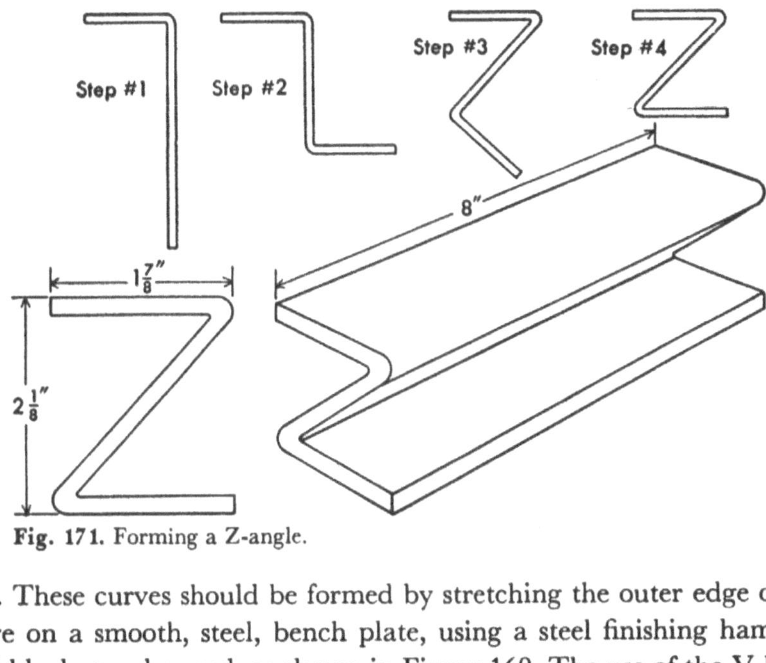

Fig. 171. Forming a Z-angle.

5 in. These curves should be formed by stretching the outer edge of the curve on a smooth, steel, bench plate, using a steel finishing hammer. A V-block may be used, as shown in Figure 169. The use of the V-block may cause buckles to appear in the material which must be hammered

out on the plate. In removing buckles, a wooden or rubber mallet should be used. In all forming operations, the greatest care must be taken to avoid dents, scratches, or other injury to the material.

Fig. 172. Joggle blocks for use in machine-forming. The pencil points to the joggle formed. (Courtesy The Glenn L. Martin Company)

Problem 5. Using 2S aluminum sheet 12 in. long and ½ in. wide, having a thickness of 0.125 in., form the material into a complete circle. Most of the bending of this material may be done by using the V-block and a soft mallet, carefully working out on a bench plate any buckles that form.

Problem 6. Form a Z-angle, as shown in Figure 171, using 24S–O alclad sheet having a thickness of 0.040 in. The material should be approximately 8 in. long and the width should be figured by making the proper

Fig. 173. Forming joggles by use of joggle blocks and press. (Courtesy The Glenn L. Martin Company)

bend allowance in order that the finished angle may have the dimensions shown in Figure 171.

Problem 7. It is often necessary in aircraft construction to form joggles. Joggles may be formed by joggle dies in a press, but it is often necessary for the sheet-metal worker to form the joggles by hand. It is not necessary to figure bend allowance for joggles, although it is good practice to allow $\frac{1}{16}$ in. The depth of the joggle is usually not greatly in excess of the thickness of the material in which the joggle is formed. The length of the joggle should not be less than about three times the thickness of the sheet. In thin, flat material, a joggle may be formed by using a joggle plate such as shown in Figure 176. The joggle is formed by clamping the material, over which the joggle plate has been slipped, in a vise.

A joggle form block may be prepared, and the joggle made by striking the metal with a cross-peen mallet forming the metal over the joggle in the block. A piece of wood having a wedge-shaped end may be used to form the joggle. This wedge is struck with a mallet.

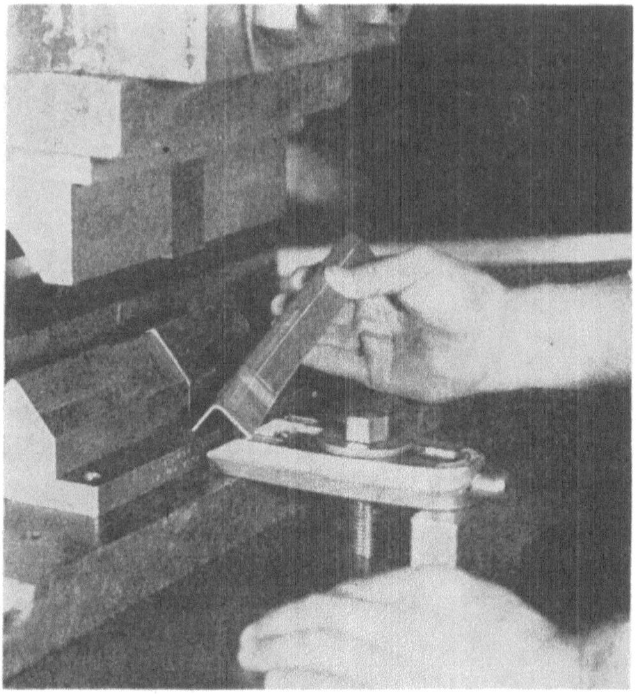

Fig. 174. Joggle formed by a joggle block and press. (Courtesy The Glenn L. Martin Company)

Problem 8. Form a joggle in a 90° flange using 24S–O sheet. Form a right-angled flange having the dimensions shown in Figure 178. Note that the joggle on one flange is the same as a joggle in a flat sheet, but on the other flange the whole width of the flange is moved sideways to form the joggle. Start the joggle across one flange, at the same time

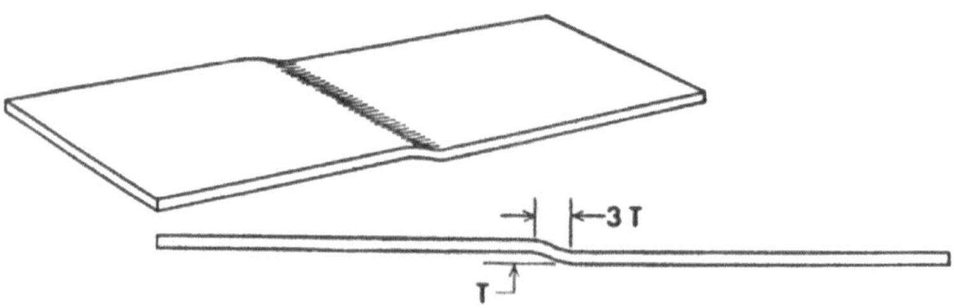

Fig. 175. A joggle formed in a flat strip.

working the material in the other flange downward over the side of the form block.

Problem 9. Form a joggle in a curved flange using 24S–O alclad having a thickness of 0.040 in. The proper form block should be prepared, and the flange clamped to the block while the joggle is being formed.

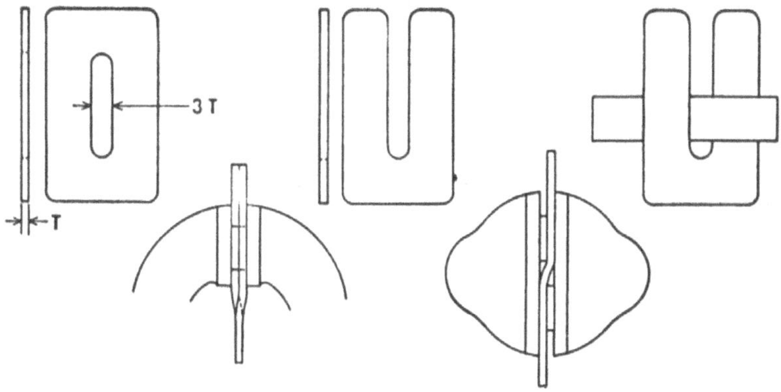

Fig. 176. Joggle plates for use on light material.

Problem 10. Form a flanged hole in a sheet of 2S aluminum 6 in. square having a thickness of 0.040 in. The diameter of the finished hole should be 3 in. with a flange ⅜ in. deep. Find the exact center of the sheet and scribe at this point with a soft lead pencil a circle having a radius of 1⅛ in. Using a small pair of curved, aviation snips, cut out the circle, leaving 1/16 in. of the material inside the pencil mark. This hole may be

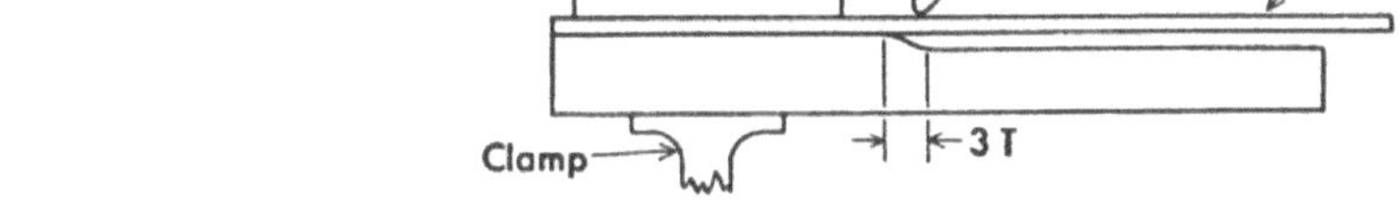

Fig. 177. A method of forming a joggle by hand.

cut out by first drilling a hole through the material well inside the circle. This hole is used as a starting point for cutting out the material, as shown in Figure 179. With a half-round vixen file, ream out the hole to the center of the pencil mark. This excess material is left after cutting with the snips so that the filing may remove all strained or torn material along the edge of the cut. Be sure that the inner edge of the hole is free from all burrs or scratches. This inner edge will have to stretch to form the flange, and any scratch or break, however small, will cause failure

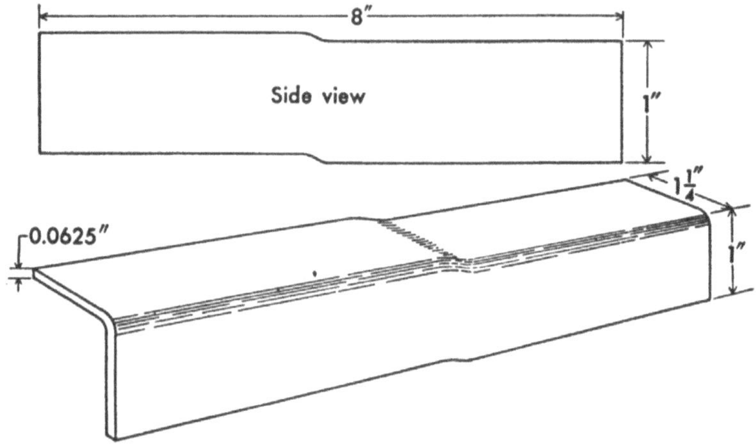

Fig. 178. Joggle formed in a right-angle flange.

in the material. The flange may be formed with a forming block and die, as shown in Figure 179, or it may be formed by using a round piece of hardwood about 5 in. long, having a diameter of approximately 3/4 in.,

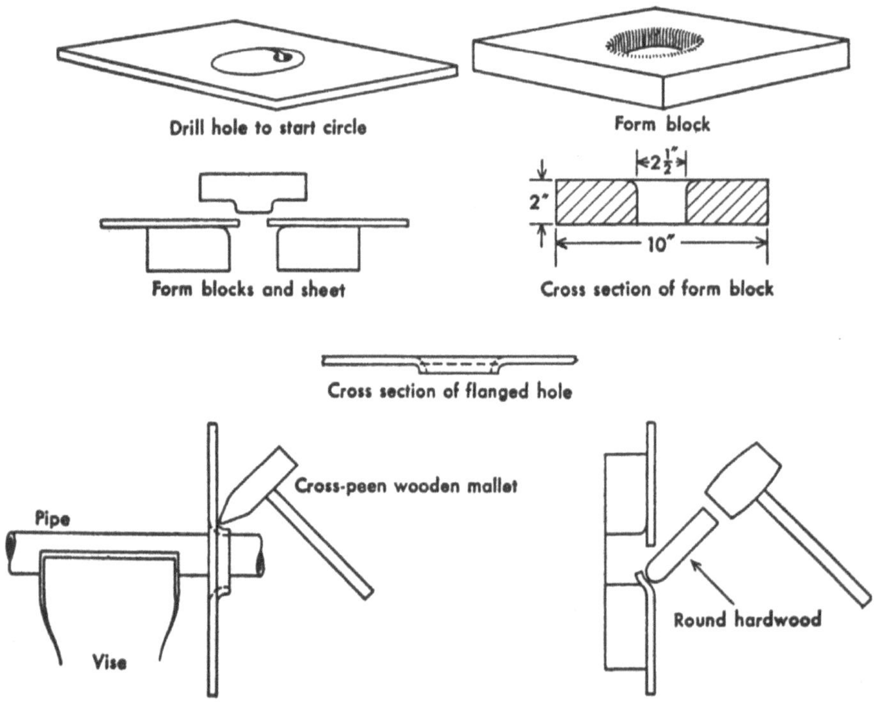

Fig. 179. Various methods of forming a flanged hole.

and rounded at one end as shown in Figure 179. The flange should be formed along the bend line first and gradually stretched into shape. The flange may also be formed, using a wedge-peen wooden mallet and a round stake or smooth piece of pipe. The radius of the flange should be continuous with only the edge of the flange perpendicular to the sheet itself.

Problem 11. Using a piece of 24S–O sheet, form a flange having a varying angle and the dimensions shown in Figure 180. One end of the flange has an angle of 60°, and the other an angle of 120°. Each side

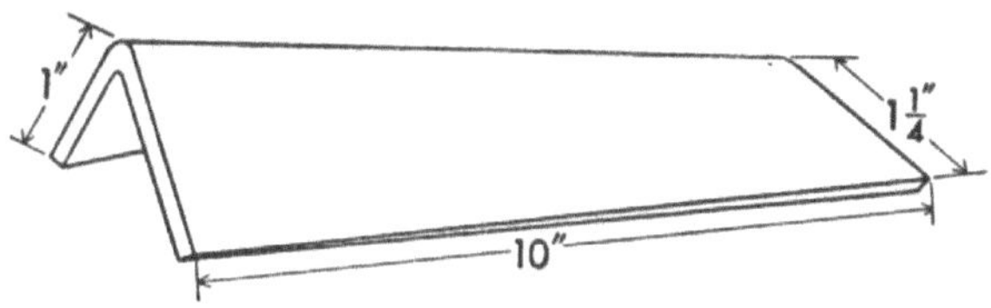

Fig. 180. A flange having a varying angle.

of the flange is bent uniformly to make the angle which tapers from end to end of the flange.

Problem 12. Form a channel of 24S–O sheet having the dimensions shown in Figure 181. In this problem, a double bend allowance must be figured. The material is clamped to the proper form block. It is better, whenever possible, to form both flanges at the same time, working each flange gradually into place.

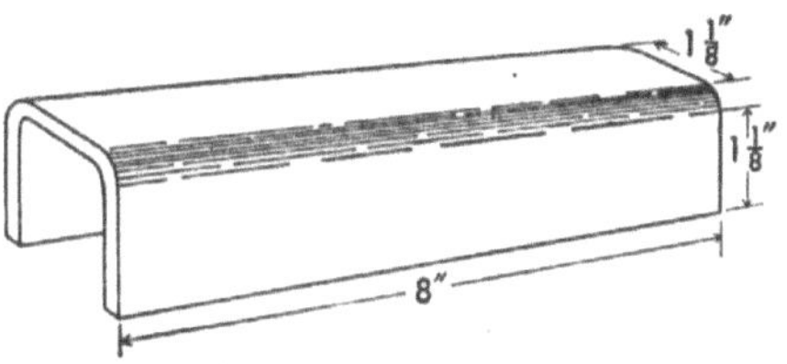

Fig. 181. Forming a channel.

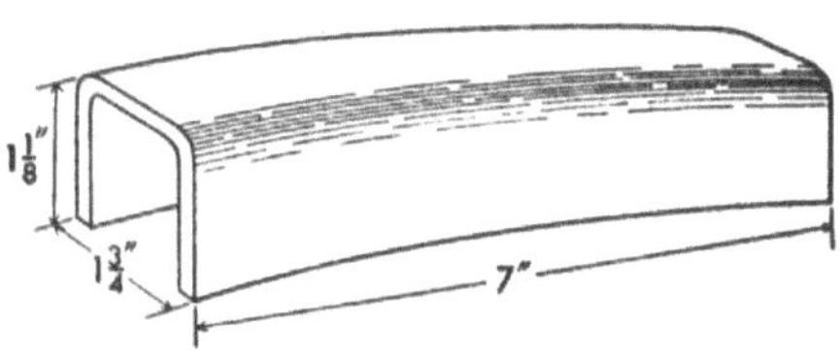

Fig. 182. Forming a curved channel.

Problem 13. Form a curved channel, as shown in Figure 182, using 24S–O alclad. This problem involves the shrinking of both flanges. The sheet should be formed to fit the top of the block and clamped firmly into place with a curved backing block. The flanges are then formed over the sides of the block. Care should be taken that severe wrinkles do not develop. A backing-up tool should be used under the edge of the flange to prevent buckling.

Problem 14. Using the curved channel formed in Problem 13, form a joggle 1 in. from each end.

Problem 15. When sheet metal is bent in such a manner that a sharp corner is formed, cracks or wrinkles almost invariably develop. Lay out the pattern for a box, as shown in Figure 184, using 24S–O sheet having a thickness of 0.063 in. The corners should be relieved by drilling

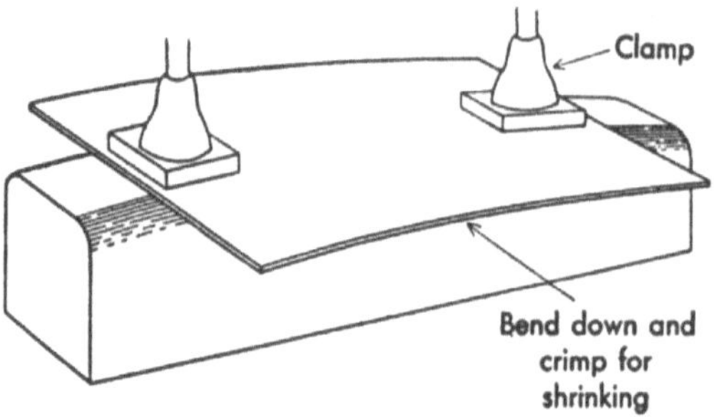

Fig. 183. The sheet clamped to a curved form block.

holes at the intersection of the lines where the corners are to be cut out, using a No. 30 drill. The corners are then cut out and the edges smoothed with a file, removing all burrs and scratches. The corners where the holes were drilled are rounded out with a small, rat-tail file. The sides are then bent over a form block, forming the sides of the box.

Problem 16. Using 24S–O sheet having a thickness of 0.040 in., form a corner reinforcement having the dimensions shown in Figure 185. The two side flanges are first formed over the corner of a form block and

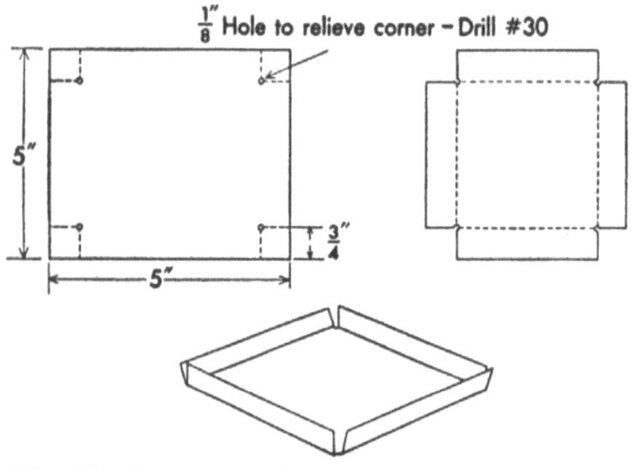

Fig. 184. A pattern to form a metal box.

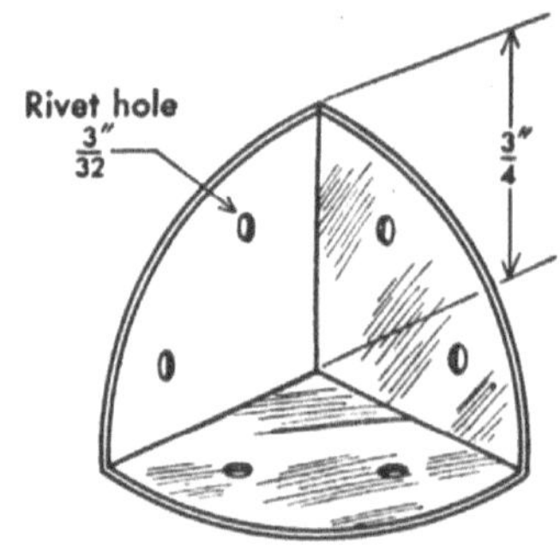

Fig. 185. Making a corner reinforcement.

clamped into place. The remaining side is then formed. Note, in forming the third side, that shrinking is involved and care must be taken to prevent wrinkles from forming.

Problem 17. Form a depression in a sheet, as shown in Figure 186. Obtain a piece of 2S sheet about 10 × 18 in., having a thickness of 0.040 in. A form block should be prepared, and the material clamped to the top of the block. Do not clamp so tightly that the material cannot

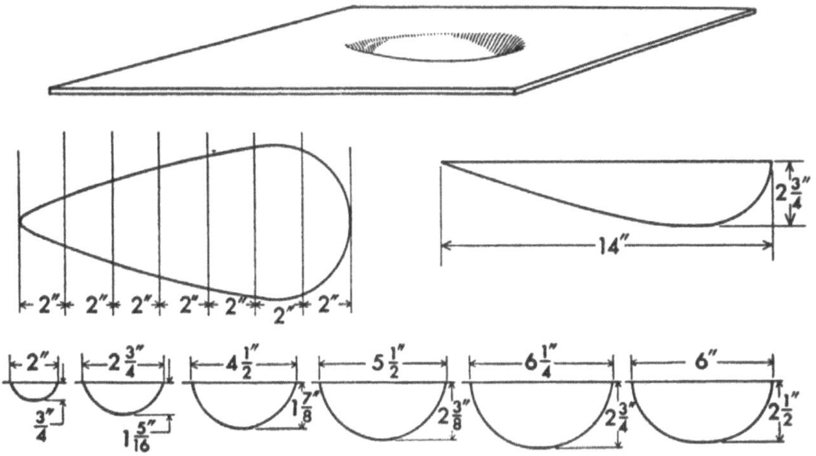

Fig. 186. A streamlined depression formed in a sheet.

slip under the clamps while being formed. With a soft, round-faced mallet, form the material into the hollow in the block. The sheet should be struck in such a manner that the material is stretched along the edge of the depression. If any wrinkles or buckles start forming along the

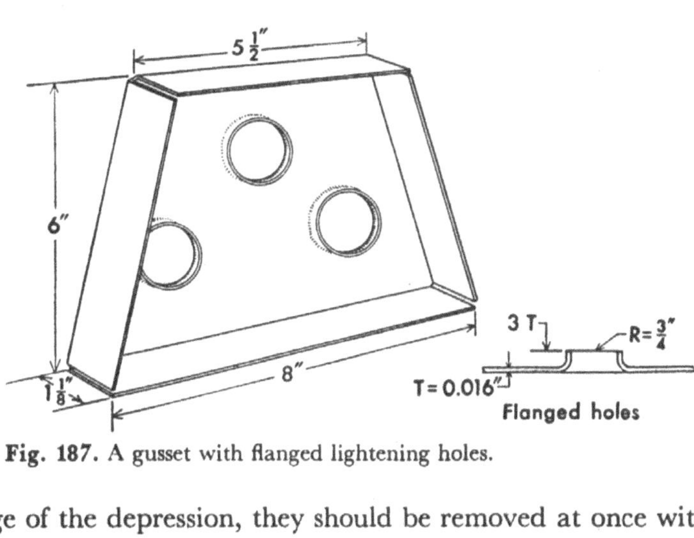

Fig. 187. A gusset with flanged lightening holes.

top edge of the depression, they should be removed at once with a flat-faced soft mallet. This is a stretching operation, and care must be taken to thin the sheet uniformly. If blows are struck on the unsupported metal in the center of the depression, the metal will stretch excessively at this point, causing thin spots.

Problem 18. Using 24S–O sheet having a thickness of 0.016 in., form the gusset shown in Figure 187. This is very thin material, and care must be used in forming the various parts not to cause excessive distortion. At the corners where the flanges are formed, holes should be drilled in the manner described in Problem 16.

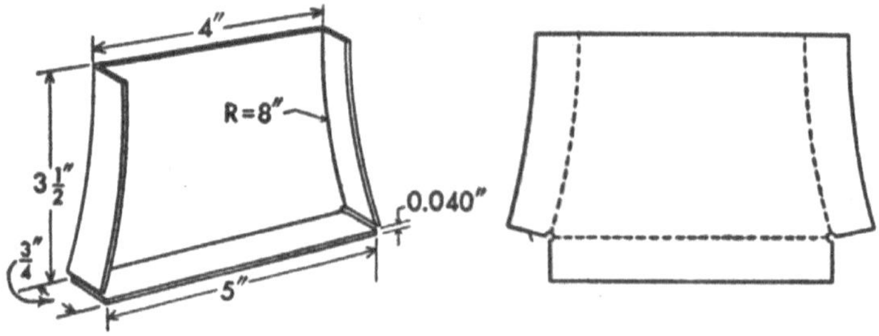

Fig. 188. A gusset with curved flanges.

Problem 19. Form a gusset having the dimensions given in Figure 188, using 17S–O sheet having a thickness of 0.040 in. This gusset has two curved flanges and one straight flange. The corners where the straight flange and the curved flange meet should be relieved by drilled holes. A proper form block should be used. The two curved flanges should be formed first.

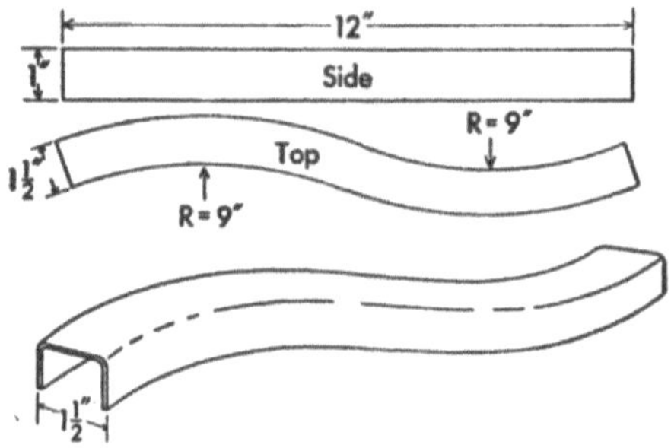

Fig. 189. A double curved channel.

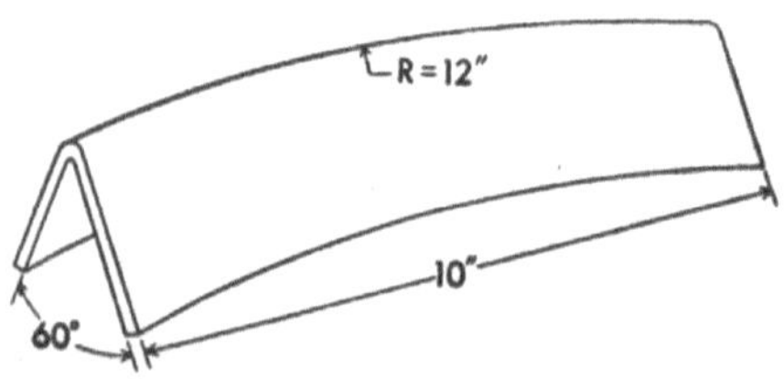

Fig. 190. A curved angle.

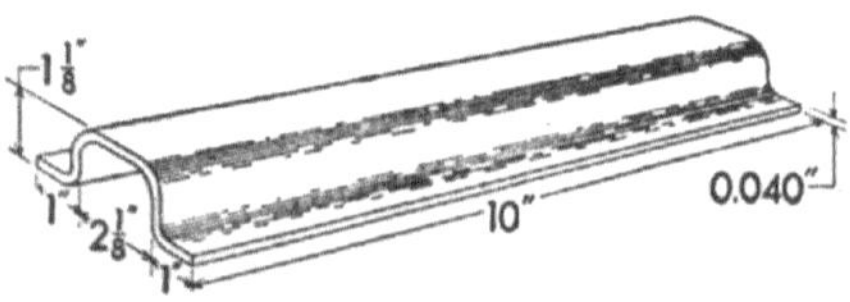

Fig. 191. A "hat" channel.

Problem 20. Using 17S–O sheet having a thickness of 0.040 in., form a double curved channel having the dimensions given in Figure 189.

Problem 21. Using 17S–O sheet having a thickness of 0.030 in., form a curved angle having the dimensions shown in Figure 190. The angle should be first formed and then shrunk over the proper form block. In this problem, both flanges must be shrunk. The shrinking should take place evenly on both sides.

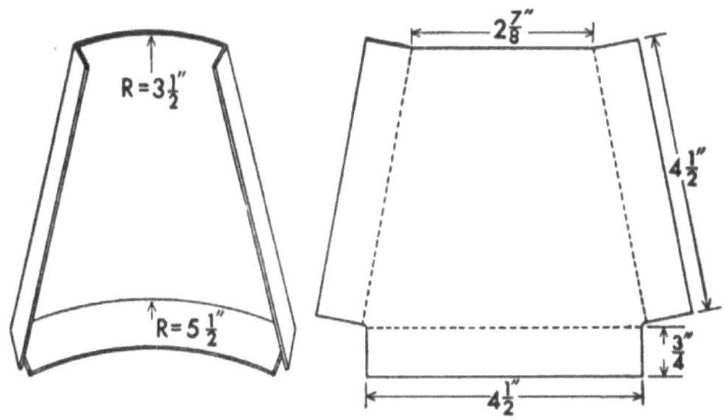

Fig. 192. A corner gusset.

Problem 22. Using 17S–O sheet having a thickness of 0.040 in., form a hat channel having the dimensions shown in Figure 191. Note that there are four bends for which allowance must be made in laying out the material to form this part.

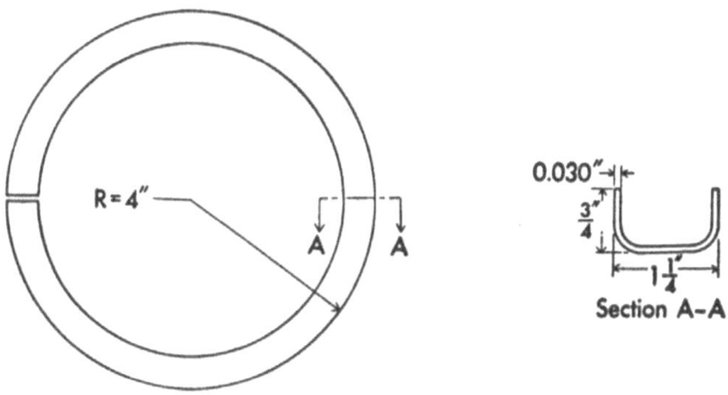

Fig. 193. A channel formed into a circle.

Problem 23. Using 24S–O sheet having a thickness of 0.063 in., form a corner gusset having the dimensions shown in Figure 192.

Problem 24. Using 17S–O sheet having a thickness of 0.030 in., form a circular channel, as shown in Figure 193. This may be done by two methods. One, by forming the sheet into a cylinder and clamping to a circular form block. The sides of the channel are then bent over the edges of the block. Two, a straight channel may be formed, and the sides crimped until a circle is formed. The channel is then clamped around a circular form block, and the crimps are removed with a soft mallet.

Problem 25. Form a fuselage nose piece from 2S sheet having a thickness of 0.016 in. and the dimensions shown in Figure 194. This part is

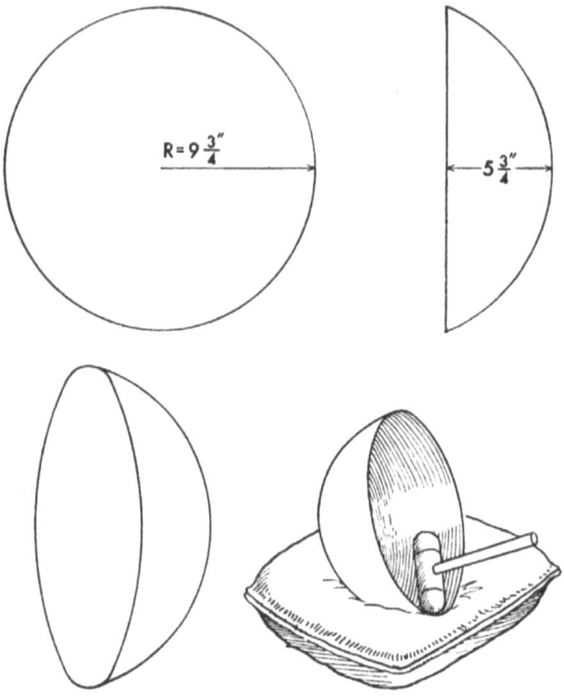

Fig. 194. Forming a nose piece with a sandbag and mallet.

to be formed with the use of a sandbag and mallet. In forming with a sandbag, a hollow is made in the top of the bag which is on a firm support. The sheet to be formed is held over the hollow and struck with a soft rawhide or rubber mallet having a slightly rounded face. The blows should be just heavy enough to bring about the desired effect on the metal. The material is quite thin, and care should be taken not to form dents which will be difficult to remove.

Problem 26. Form a tail piece for a sheet-metal rib from 24S–O sheet having a thickness of 0.030 in. and the dimensions shown in Figure 195.

Problem 27. Form a nose piece for a sheet-metal rib, using 24S–O sheet having a thickness of 0.030 in. and the dimensions shown in Figure 195.

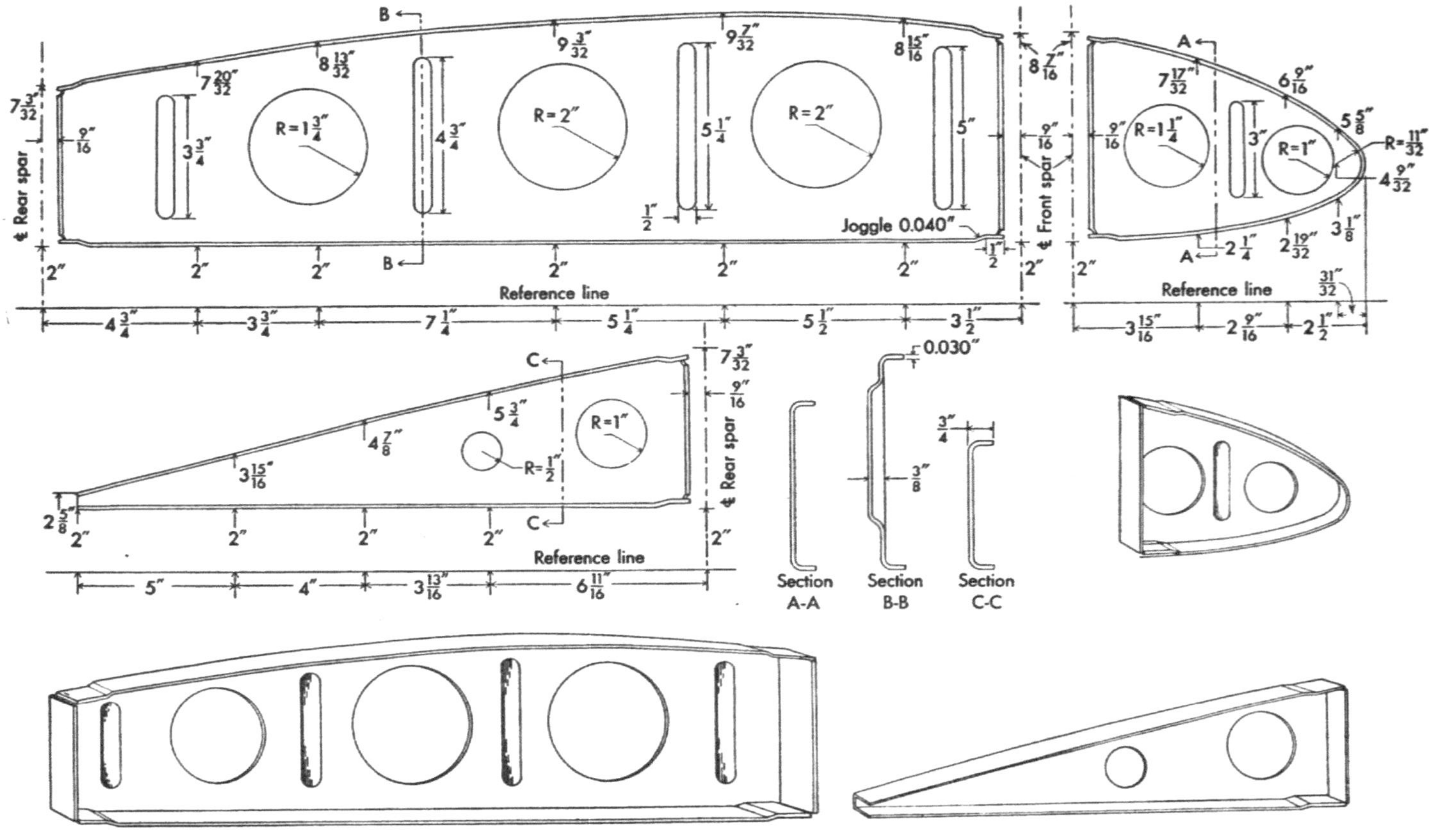

Fig. 195. Parts of a sheet-metal rib.

Problem 28. Form a web for a sheet-metal rib, using 24S–O sheet having a thickness of 0.030 in. and the dimensions shown in Figure 195.

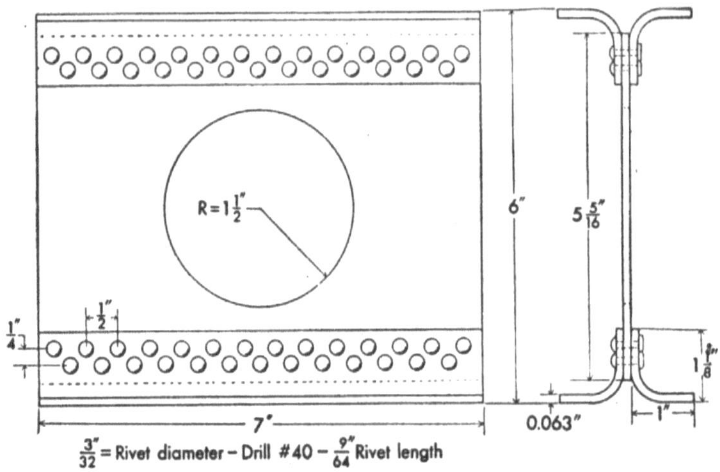

Fig. 196. A section of a sheet-metal spar.

Problem 29. Form a section of a sheet-metal spar, using 24S–O sheet having a thickness of 0.063 in. and the dimensions shown in Figure 196.

XVII DRILLING

Drilling is an important part of sheet-metal work. It is necessary for the sheet-metal worker to be able to drill holes accurately for rivets and other types of fasteners. Drilling requires a considerable degree of skill.

Most drilling in metal is done with a twist drill. As shown in Figure 197, the drill is made up of three principal parts: the point, the body, and the shank. Two spiral grooves run along opposite sides of the body. These grooves or flutes help form the cutting edges on the point. They also assist in removing the chips and allow lubricants to penetrate to the point of the drill. For general use, the point is usually ground to an angle of approximately 59°. This angle varies with the type of material being drilled. The dead center is located at the center of the point and should be at the exact center of the drill, if the drill has been properly sharpened. The point is the cone-shaped end of the drill. The edges of the flutes form the cutting part of the drill, or the lips. After sharpening, the lips should be equal in length, and the angle should increase somewhat to give lip clearance and allow the cutting lip to bite into the material. The angle of lip clearance should be gradually increased as the center of the drill is approached. The line across the dead center of the drill should form an angle of about 135°. Improper grinding is a common cause for broken drills. On the edge of the flute is a small, narrow part which is equal to the diameter and allows proper clearance in the hole, preventing the drill from binding. This is called body clearance. The metal column which separates the flutes is called the web and forms the main supporting column of the drill. The web thickens from the point to the shank.

The part of the drill which fits into the chuck is called the shank. Drills having a tapered shank for use in drill presses or machine drilling usually have a tang at the end of the shank which fits in a slot to prevent the drill turning in the chuck.

The successful use of drills depends largely upon their proper grinding. Two of the common faults in grinding drills are unequal length of the lips and an improper angle of lip clearance. The metal back of the lip itself should be ground away at an angle of about 12°. If this metal is not ground away, the lip has no clearance. Without clearance, the drill will not bite into the metal, but will slide over the surface without cutting. Figure 204 shows proper lip clearance. Too much clearance will thin the lip to such an extent that the lip will chip. Chipping along the entire edge of the lip indicates too much lip clearance. Excessive speed may cause chipping of the lip toward the circumference of the drill, as shown in Figure 206. Figure 207 shows the effect of grinding one lip longer than the other or at a different angle. This will cause excessive strain on both the drill and the drill

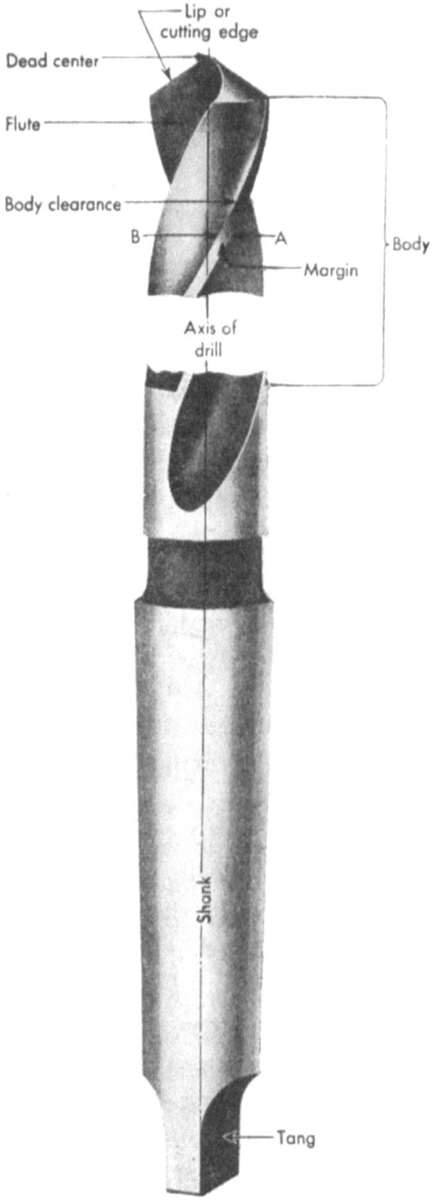

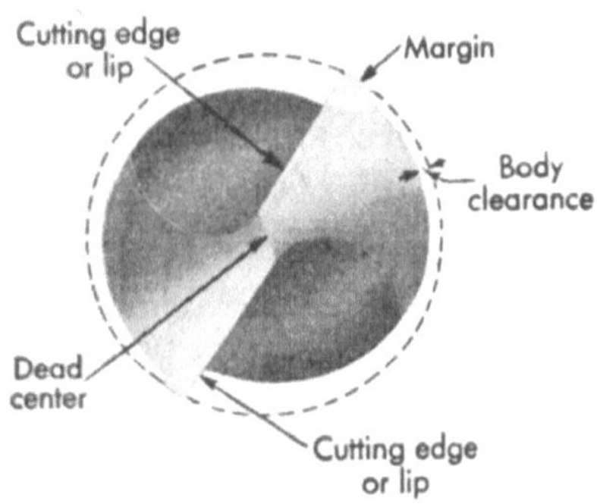

Fig. 197. Parts of a twist drill. (Courtesy The Cleveland Twist Drill Co.)

press. The hole will be uneven, scored, and larger than the diameter of the drill.

The feed of a drill is the distance that the drill moves down into the material with each revolution. The feed should be just great enough to allow the drill to cut cleanly and smoothly. If the feed is too slight, the drill may have a tendency, particularly in hard material, to ride over the surface at the bottom of the hole. This will cause the material to become hard and polished, and the drill will stop cutting. If the feed is too rapid, the drill may bite in and break. In hard material, excessive feed may cause the drill to split up the center, as shown in Figure 208.

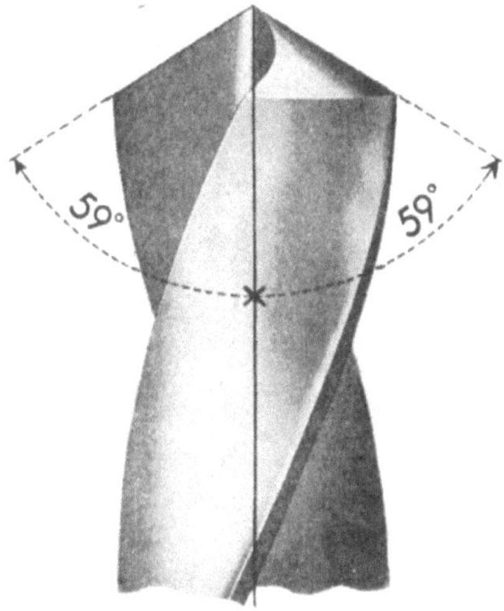

Fig. 198. Correctly ground lips. The two lips of this drill are of the same length and of the same angle to the axis of the drill. (Courtesy The Cleveland Twist Drill Co.)

The rake angle of a drill is built into the drill and is the angle between the edge of the flute and the surface of the work. This angle causes the chips to curl and clear the hole. The greater the angle, the tighter the chips tend to curl. The rake angle built into the drill by the manufacturer should not, as a rule, be changed. The type of material being drilled determines the rake angle. If it is necessary to increase this angle, the lip may be ground away on the inside of the flute. This might be done when drilling harder material than that for which the drill was designed.

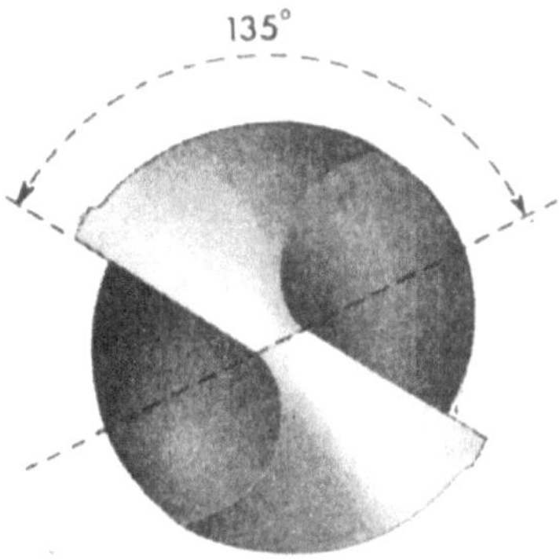

Fig. 199. How to gauge the correctness of lip clearance angle. (Courtesy The Cleveland Twist Drill Co.)

Fig. 200. Lip clearance. The drill on the left has no lip clearance. The drill on the right is properly cleared. (Courtesy The Cleveland Twist Drill Co.)

As the drill wears away, due to many sharpenings, it may be necessary to thin the web. This may be done by grinding it with a round-faced emery wheel or by grinding it with a flat, sharp-cornered emery wheel which produces a notched point, as shown in Figure 210.

Drill sizes are designated as follows: by numbers from 80 to 1, corresponding in drill sizes from 0.0135 to 0.228; by the letters, A to Z, which correspond to the drill sizes, 0.234 to 0.413; and by fractions from 1/64 in. to 4 in. and over, by sixty-fourths. Drill sizes are also given in the metric system

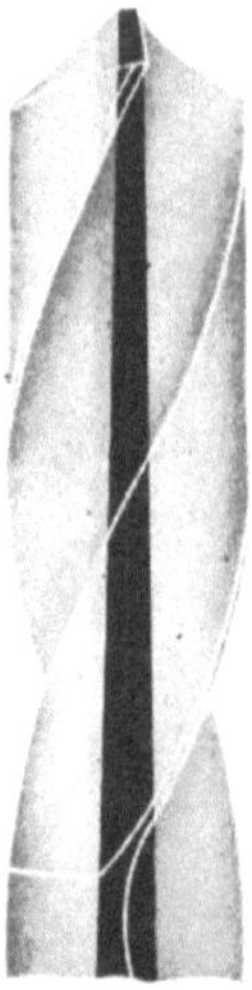

Fig. 201. The web is the metal column which separates the flutes and is the supporting section of the drill. (Courtesy The Cleveland Twist Drill Co.)

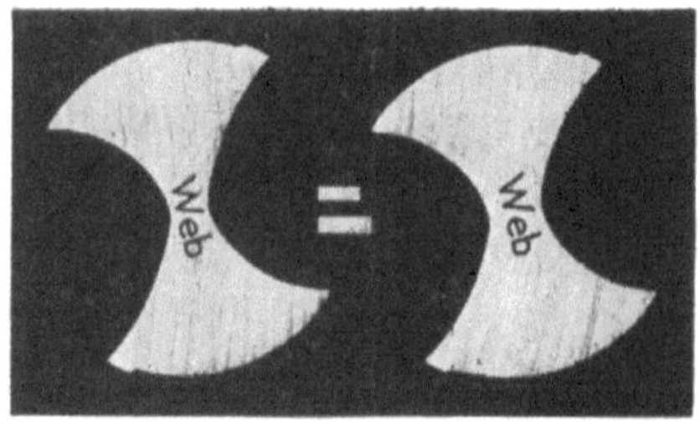

Fig. 202. Sectional view of the web. The section on the left was cut from a drill near the point, while the section on the right was cut near the shank. The difference in the thickness of the web at these points is shown by the length of the white lines between the two sections in the illustration. (Courtesy The Cleveland Twist Drill Co.)

TABLE VI. SUGGESTED SPEEDS FOR HIGH-SPEED DRILLS

	SPEED IN F.P.M.
Aluminum and its alloys	200–300
Bakelite	100–150
Brass and bronze, ordinary	200–300
Bronze, high tensile	70–100
Cast iron, soft	100–150
Cast iron, hard	70–100
Cast iron, chilled	30– 40
Malleable iron	80– 90
Magnesium and its alloys	250–400
Monel metal	40– 50
Steel machinery	80–110
Steel, annealed	60– 70
Steel, tool	50– 60
Steel forgings	50– 60
Steel, alloy	50– 70
Steel, stainless, free machining	60– 70
Steel, stainless, hard	30– 40
Slate, marble and stone	15– 25
Wood	300–400

Fig. 203. A large drill press. Note shank of drill inserted in the chuck. (Courtesy Douglas Aircraft Company, Inc.)

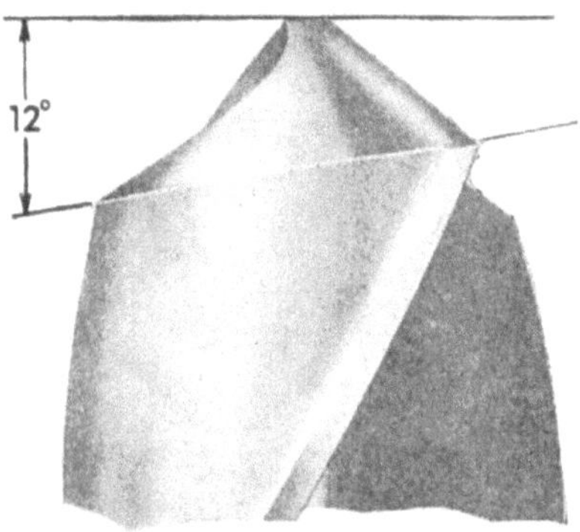

Fig. 204. The proper way to grind the surface back of the cutting lip. The angle indicated is at the circumference of the drill. This angle should be increased as the center of the drill is approached. (Courtesy The Cleveland Twist Drill Co.)

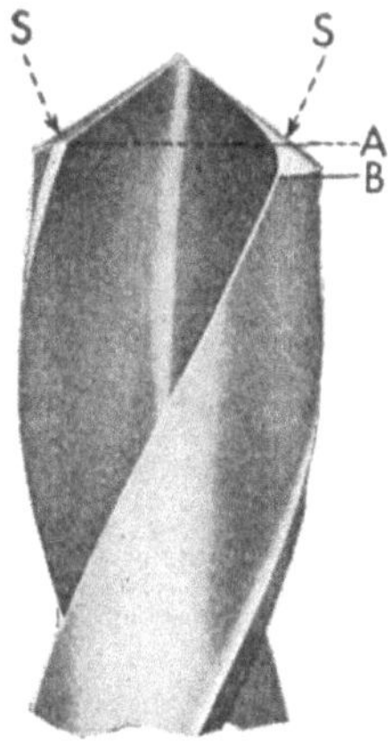

Fig. 205. Proper lip clearance. Note how much lower the heel line "B" is than the cutting lip line "A". This difference is the measure of clearance. (Courtesy The Cleveland Twist Drill Co.)

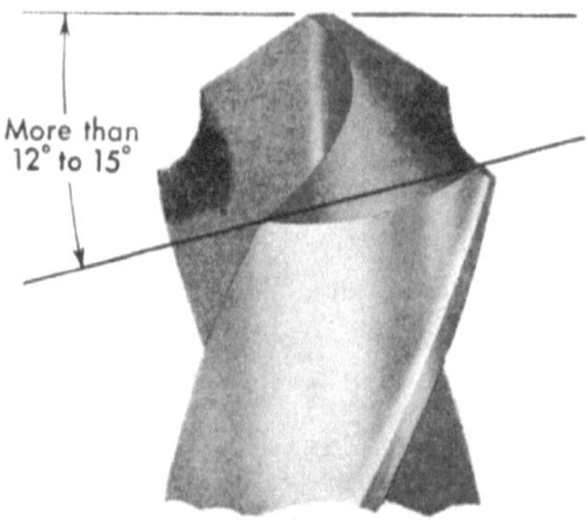

Fig. 206. Results of giving a drill too great lip clearance, or operating at an excessive speed. (Courtesy The Cleveland Twist Drill Co.)

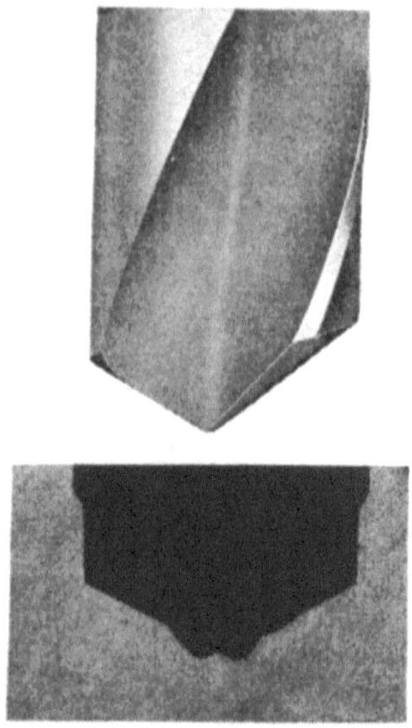

Fig. 207. The angles of the lips are of different lengths. Note the effect on the hole. (Courtesy The Cleveland Twist Drill Co.)

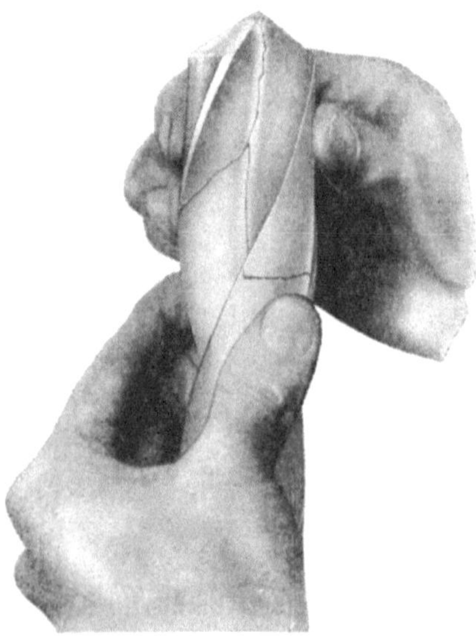

Fig. 208. This drill does not have enough lip clearance. As a result, cutting stopped and as the feed pressure increased, the drill "split up the center." (Courtesy The Cleveland Twist Drill Co.)

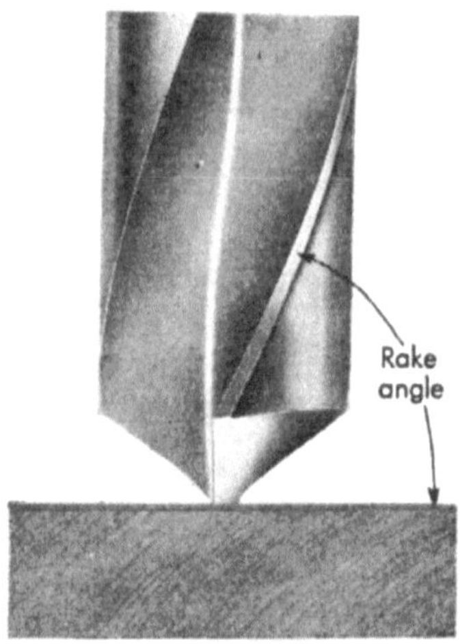

Fig. 209. The rake angle is the angle of the flute in relation to the work. (Courtesy The Cleveland Twist Drill Co.)

in millimeters from 0.5 to 10 mm., by 1/10 mm. Above 10 mm., the size increases by 0.5 mm. A millimeter is approximately 1/25 in. and 1/10 mm. is approximately 1/250 in.

The speed of a drill is the speed of its circumference or the distance that the drill would roll over a flat surface. For example, if the drill

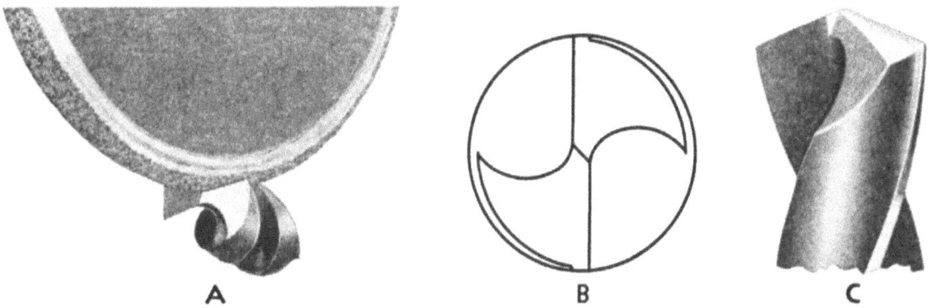

Fig. 210. Two methods of thinning the web of a drill. "A" using a round-faced emery wheel; "B" the result of using a round-faced emery wheel. "C" a notched point formed by a sharp-cornered emery wheel. (Courtesy The Cleveland Twist Drill Co.)

speed is 300 ft. per min., the drill would roll along a flat surface 300 ft. in one minute. The speed of a drill varies with the kind of material being drilled.

It is necessary at times to use a lubricant to assist in cutting. For brass and for aluminum and its alloys, gasoline and lead-oil compounds may be used. Cast iron is usually drilled dry; malleable iron, Monel metal, and steel may be drilled with a good nonviscous, neutral oil. When drilling hard material, turpentine is sometimes used to assist the lips in biting into the material.

The common sizes of rivets and the corresponding drill sizes are given in Table VII.

TABLE VII. COMMONLY USED RIVET AND DRILL SIZES

RIVET	DRILL	
1/16	#51	(0.067)
3/32	#40	(0.098)
1/8	#30	(0.128)
5/32	#20	(0.161)
3/16	#10	(0.193)
1/4	F	(0.257)
5/16	P	(0.323)
3/8	W	(0.386)

A chuck key should always be used in inserting or removing drills. Be sure that the right size of drill has been obtained. The drill should be sharp and properly ground. Remove dirt, oil, and chips from the drill chuck and parts of the motor. Place the drill motor on a table or bench when inserting or removing drills. With the chuck key, open the jaws only far enough to admit the shank of the drill. Insert the drill all the way into the chuck and tighten the chuck firmly with the key. Be sure to remove the chuck key before starting the motor. Be sure, also, that the switch on the motor is off before plugging in the cord.

Fig. 211. A multiple drill press. Note that the work is fastened firmly in place. (Courtesy Douglas Aircraft Company, Inc.)

Suitable goggles should always be worn when drilling. The material being drilled should be well supported, as sagging or bending of the material will cause the drill to break. A vise may be used to hold small parts while drilling. Large parts may be supported with blocks and clamps.

It is necessary that drilled holes be accurately located. Hold the motor with one hand, grasping the drill chuck between the thumb and forefinger of the other hand. Rest the back of the hand or knuckles against the material to be drilled. This will give steady support and allow the drill point to be accurately centered. Turn the chuck lightly with the fingers to make a slight depression in the metal. Examine this depression to see that it is at the exact point where the hole is to be drilled. Do not attempt to drill the hole if the mark is not exactly centered. If the mark is in the proper location, which should be determined by crossed lines, hold the front of the drill motor in the palm of the hand, resting the thumb and forefinger against the work. Be sure that the drill is at right angles to the sheet before starting the motor. Supporting the motor in this manner prevents the drill chuck from injuring the sheet as the drill penetrates the metal. Be sure the chuck does not come in contact

Fig. 212. An extension drill inserted in a chuck. (Courtesy Douglas Aircraft Company, Inc.)

with the sheet as the hole is finished. It may be well to place a fiber washer over the drill to protect the sheet from injury when a short drill is being used.

The motor should continue to turn while the drill is withdrawn from the hole. Keep the drill at right angles to the sheet when removing from the hole. The point of the drill should penetrate through the material approximately ¼ in. If a drill is held at a wrong angle, it will cause elongated holes and may damage the surrounding parts. Turn off the motor before laying down the drill. Always check to see that the hole is properly centered. If the hole is not properly centered, notify the person in charge.

It is sometimes necessary to use an extension drill. An extension drill may have a long shank or be made by fastening the proper-sized drill into the end of a steel rod. Always support the extension drill near its

Fig. 213. Drilling holes located by means of templates. (Courtesy The Glenn L. Martin Company)

end before starting the motor, as the drill may whip, causing it to break. Select a drill just long enough to penetrate the work. Out-of-the-way places may be reached by using a snake drill which has the chuck at the end of a flexible cable. It may be necessary for two persons to operate this type of drill, one holding the snake and the other the motor. A number of attachments may be obtained which allow holes to be drilled at various angles and in out-of-the-way places.

Pilot holes are small holes drilled to locate the exact centers of the finished holes. These holes are drilled to full size after the parts have been assembled, before inserting the fasteners.

In aircraft construction it is often necessary to locate holes by means of drilling jigs or templates. When drilling through a jig or template, the drill should be inserted before the motor is started. Care should be taken not to injure the jig or template.

When laying out material for the location of holes to be drilled, unless specifically allowed, all marks should be made with a soft, sharp pencil. If laid out with a steel scriber, the scratches are almost sure to cause fractures in the material later on. Aluminum-coated alloys and material which has been painted or has had other protective treatment should never be marked with any but a soft pencil. Before drilling, the center of the hole may be marked with a prick punch or center punch. This will help to start the hole in the proper place. The material should always be backed up by a smooth, solid surface to prevent denting and marring. When drilling, the point of the drill should be placed in the depression made by the punch. When drilling hard material, such as stainless steel, the drill has a tendency to jump out of the depression unless held firmly. It is difficult to start a hole in hard material with the drill without first center punching.

If the material being drilled is thick, it may be necessary to remove the drill from time to time to clear the hole of chips. This is particularly true when drilling wood, fiber, or other soft material. When drilling hard material, it is always well to use a lubricant. When drilling through two or more thicknesses of material, chips are apt to gather between the sheets. These should be removed before fasteners are installed. When removing burrs from drilled holes with an over-sized drill or burring tool, care should be taken to prevent countersinking. When drilling holes in sheet by duplicating the holes in the structural framework, the sheet should always be backed up with a block of hardwood. This holds the skin in place and prevents the drill from breaking too far through the material.

XVIII RIVETING

Most modern aircraft have much of their structure fastened together by rivets. Some of the large bombers contain more than a quarter-million rivets. Rivets are used to fasten the skin in place and to join many of the structural parts.

There are four main classifications of rivets, according to head shape: round head, brazier head, flat head, and countersunk or flush head, as shown in Figure 215. The type of head used depends upon the part of the airplane that is being riveted. The countersunk head or flush type is used largely on exposed surfaces to reduce air resistance. Sometimes, on exposed surfaces, brazier-head rivets or shallow brazier-head rivets are used in place of the countersunk type. Rivets having round heads are used on structural parts. Flat-head rivets are used only on internal parts.

Rivets are also classified according to their alloy and whether or not they are driven "as received" or must be heat-treated. Rivets of 2S and 3S aluminum are quite soft and are driven cold, as received. Storing for long periods of time does not affect their driving characteristics or strength. Rivets of these alloys are used for nonstructural parts where great strength is not required.

Rivets of the alloy, 17S, are commonly used. These rivets are always driven in the fully heat-treated condition. If they are received in the 17S–O condition, they must be heat-treated and, if received in the 17S–T condition, they must be reheat-treated before driving. The rivets are heated to a temperature of approximately 940° F. and then quenched in cold water. The temperature must be accurately controlled, and these rivets should be used within 15 or 20 min. after heat treating. If not used immediately, they should be stored in a refrigerator to prevent hardening. This type of rivet is sometimes called an "ice-box rivet." These rivets reach their full strength and hardness in about four days after driving.

Rivets of the alloy, 24S–T, have a higher strength than the 17S–T rivet and also require heat treating. They harden rapidly and should

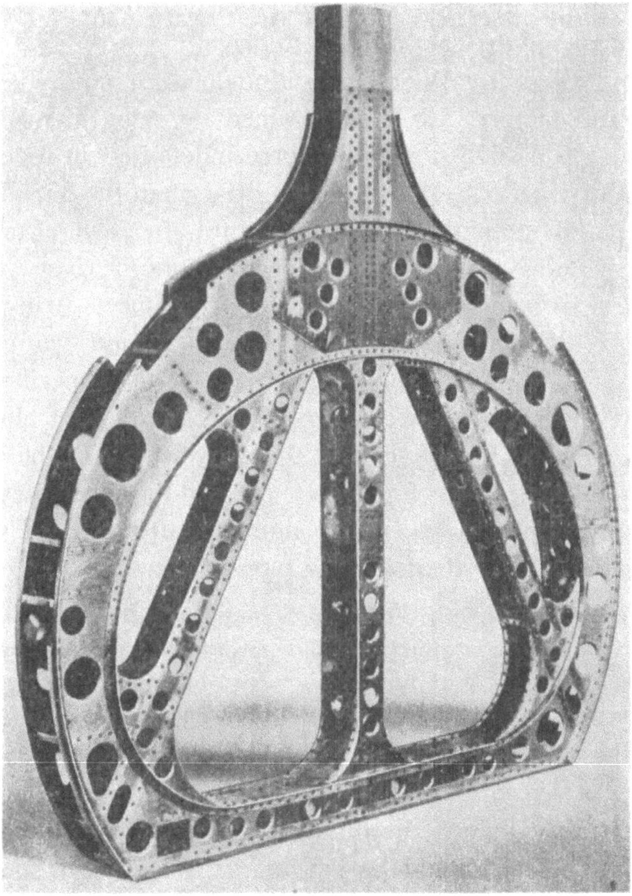

Fig. 214. A bulkhead showing large number of rivets. (Courtesy Aluminum Company of America)

be driven within a few minutes after quenching. If not used immediately, they should be stored in dry ice which will keep them soft for a period of

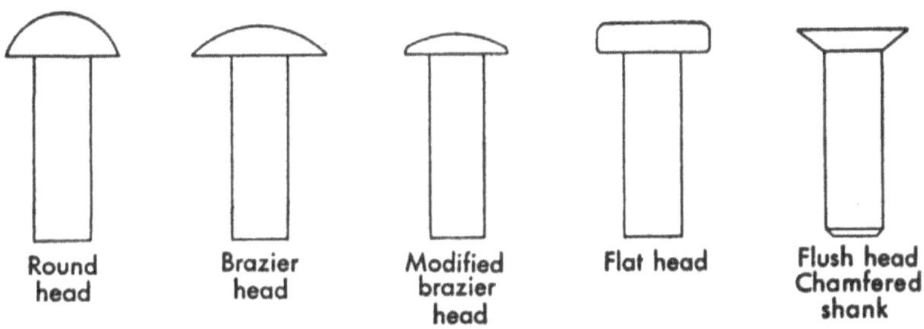

Fig. 215. Different types of rivet heads.

several weeks if held at a temperature of about $-50°$ F. This type of rivet is usually ordered in the fully heat-treated condition to guard against use in the annealed condition. After heat treating, if these rivets are allowed to age they are difficult to drive.

The alloy, 53S, in the W or T condition does not require heat treating in the aircraft factory. 53S–T is obtained by aging 53S–W for about six hours at 350° F. 53S–T may be stored indefinitely at room temperature before driving. 53S–W is easier to drive than the 53S–T, but does not reach its full strength and hardness until after several months.

The Army and Navy have set up specifications for rivets. The letters, AN, before a rivet number mean that the rivet meets Army and Navy specifications. Rivets are designated by letters and figures, such as AN425–DD4–6. The AN signifies Army and Navy specification; 425 means that the rivet has a countersunk head and is of aluminum or aluminum alloy; DD indicates that the alloy is 24S–T; the 4 gives the diameter in thirty-seconds of an inch ($\frac{4}{32}$); and the 6 indicates the length in sixteenths of an inch ($\frac{6}{16}$). The numbers after the AN which give the head style indicate the following types of rivets:

420 — An iron or copper rivet having a countersunk head.
425 — An aluminum or aluminum alloy rivet having a countersunk head.
430 — An aluminum or aluminum alloy rivet having a round head.
435 — An iron or copper rivet having a round head.
442 — An aluminum or aluminum alloy rivet having a flat head.
455 — A brazier-head rivet.

If the letter, A, follows the number given above, it indicates 2S or 3S aluminum which should be driven "as is" without heat treating. The heads on these rivets are smooth. AD following the number indicates an A17S–T alloy to be driven as is without heat treating. This rivet has a dimple in the head. D indicates a 17S alloy, ice-box rivet which should be heat-treated before driving and has a small projection on the head. DD indicates a 24S alloy, ice-box rivet which must be heat-treated before driving. This rivet has two raised bars on the head. These markings are shown in Figure 216.

The proper size of the rivet to be used is decided upon by the manufacturer and, unless permission has been received to do otherwise, the aircraft worker should use the specified rivet.

The length of all rivets except the flush-head type is measured from the lower side of the head to the end of the shank. Flush-type rivets are measured from the top of the head to the end of the shank. Some rivets have the end of the shank slightly beveled to increase the ease with which they are inserted into the rivet hole. Rivets of this type are known as chamfered-shank rivets.

As a rule, the diameter of the rivet shank should not be greater than 2½ to 3 times the thickness of the sheet to be riveted. The rivet shank

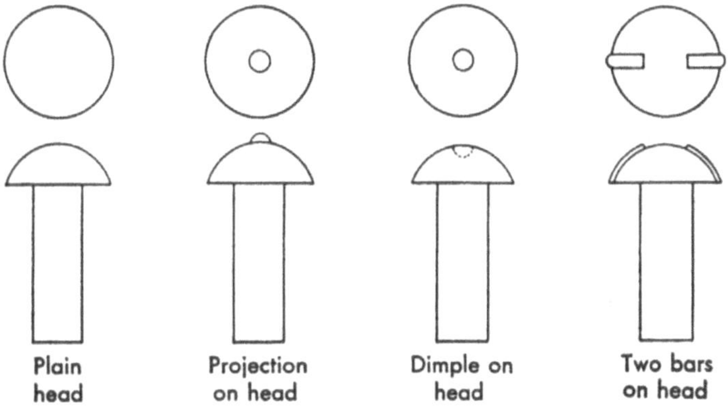

Fig. 216. Head markings for rivets of various alloys.

should not be less in diameter than the thickness of the sheet. The distance between rivets in the sheet varies with the thickness of the sheet and the required strength of the joint. Rivets are not usually placed closer together than 3 times the diameter of the rivet. A distance of not less than twice the diameter of the rivet should be allowed from the center of the rivet hole to the edge of a sheet. This is known as edge

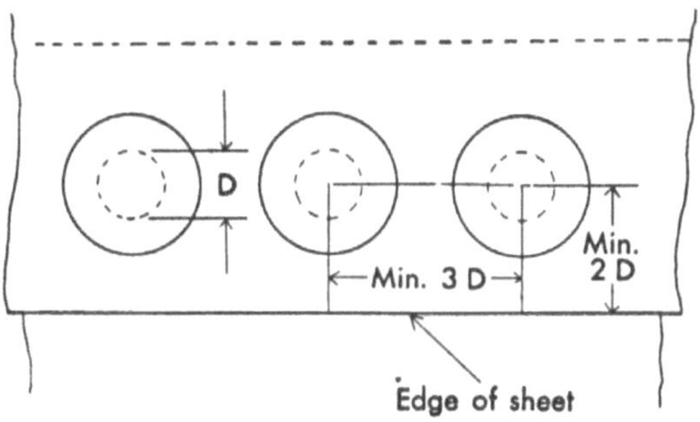

Fig. 217. Rivet spacing and edge distance.

distance. If rivets are placed too close together or too near the edge of a sheet, the sheet is apt to buckle when the rivet is driven.

Rivet holes may be drilled or punched either before or after the part is in place. There should be just enough space between the rivet and the edge of the hole to allow the rivet to slip easily into place. The sheet-metal worker should be sure that he is using the proper-sized drill before drilling rivet holes. The most common sizes for aircraft rivets are $\frac{1}{16}$, $\frac{3}{32}$, $\frac{1}{8}$, $\frac{5}{32}$, and $\frac{3}{16}$. The corresponding drill sizes for these rivets are: 51, 40, 30, 20, and 10, respectively.

The process of riveting consists of placing the rivet through matching holes in two or more sheets and forming a head on the shank end of the

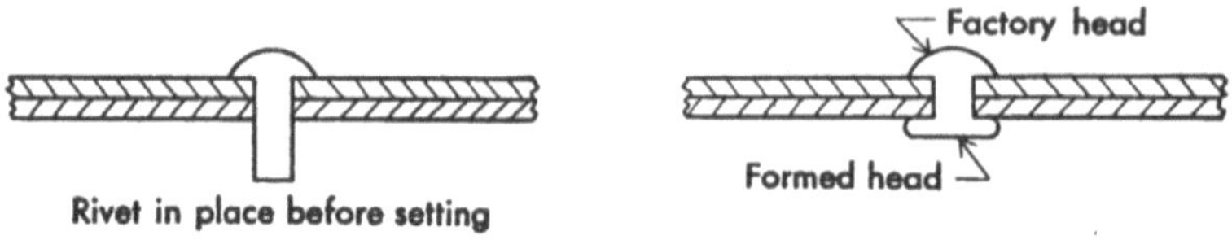

Fig. 218. A rivet placed through two sheets, showing the factory head and the formed head.

rivet to hold it firmly in place. The head already on the rivet is called the "factory head," and the head which is formed during the driving of the rivet is called the "formed head." The formed head is made by upsetting the metal in the rivet. The metal is squeezed together, shortening the shank. Since none of the material in the rivet is removed, the exposed part of the shank swells, forming the head. The unexposed part of the shank also swells, filling the hole completely and causing the rivet to fit tightly in the hole. This is called the "rivet grip." The head

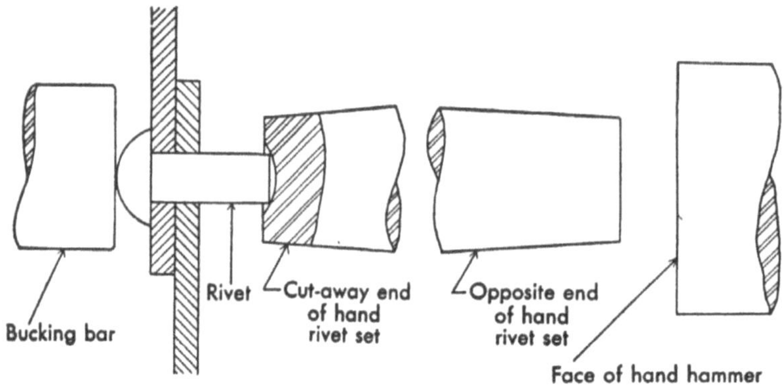

Fig. 219. A rivet being set by hand.

on the shank end is formed by squeezing the rivet between the rivet hammer and a bucking bar. A bucking bar is made of steel and must be of sufficient weight to prevent bouncing when the rivet is driven. The rivet is usually driven by a rivet gun or hammer operated by compressed air. An air- or hand-operated rivet squeeze may be used.

In driving rivets by hand, the rivet set, which is the steel tool placed against the rivet to receive the blows of the hammer, is placed against the shank end of the rivet, and the bucking bar is held against the factory head. When using a rivet gun, the rivet set is placed against the factory head, and the bucking bar against the shank. In squeeze riveting, one jaw of the rivet squeeze is placed on the factory head and the other against the shank, and the rivet is set by squeezing it into place.

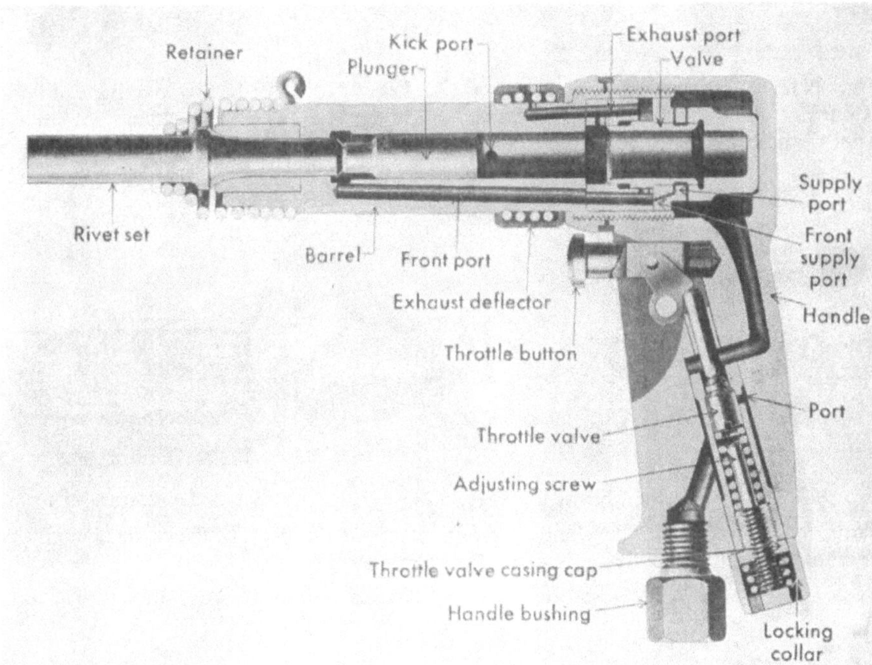

Fig. 220. The parts of a riveting hammer. (Courtesy The Cleveland Pneumatic Tool Company)

A one-shot gun strikes only one blow each time the trigger is pulled. The pressure in this gun should be so adjusted that the rivet is completely driven by one blow. The fast-hitting, or vibrating, hammer continues to strike blows at a rate of about 5000 blows per minute as long as the trigger is held back. The slow-hitting hammer strikes the blows at a much slower rate of speed, but continues to strike as long as the trigger is depressed. The speed of this hammer ranges up to 2500 blows per minute. Squeeze-rivet sets and rivet guns may be obtained in many sizes and shapes for different types of work, as shown in Figures 221, 222, 223, 224, 225, 226, and 227.

The length of the rivet should be such that the shank extends through the material being riveted 1½ times the diameter of the rivet. The thickness of the formed head should be ½ the diameter of the rivet after forming.

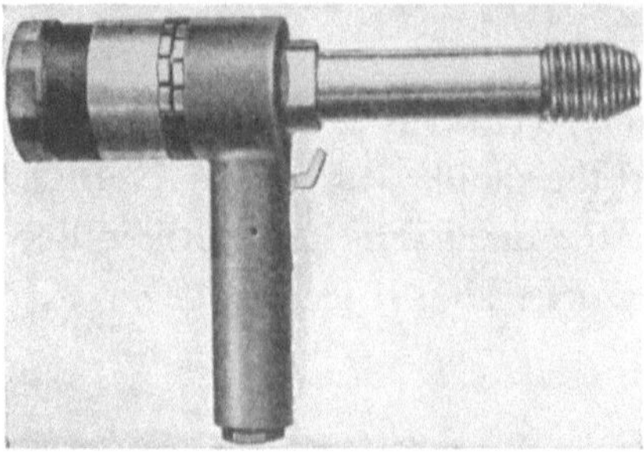

Fig. 221. A one-shot riveting hammer. (Courtesy The Cleveland Pneumatic Tool Company)

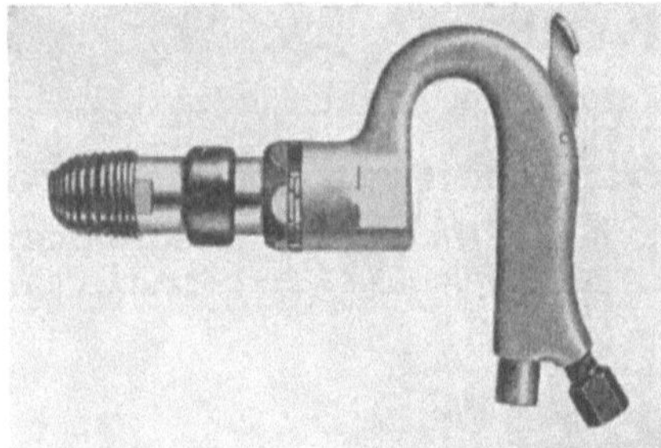

Fig. 222. A fast-hitting riveting hammer. (Courtesy The Cleveland Pneumatic Tool Company)

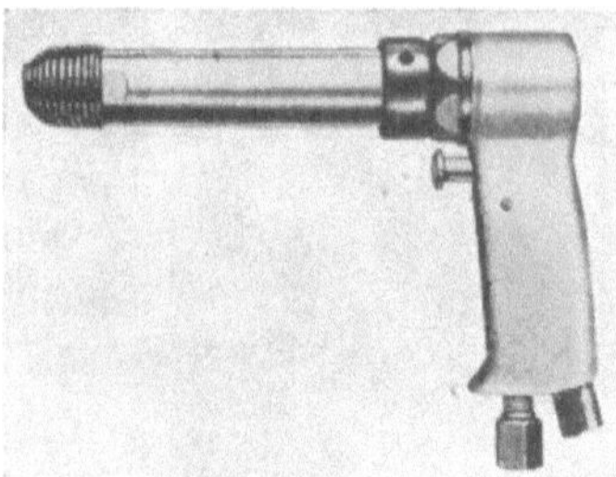

Fig. 223. A slow-hitting riveting hammer. (Courtesy The Cleveland Pneumatic Tool Company)

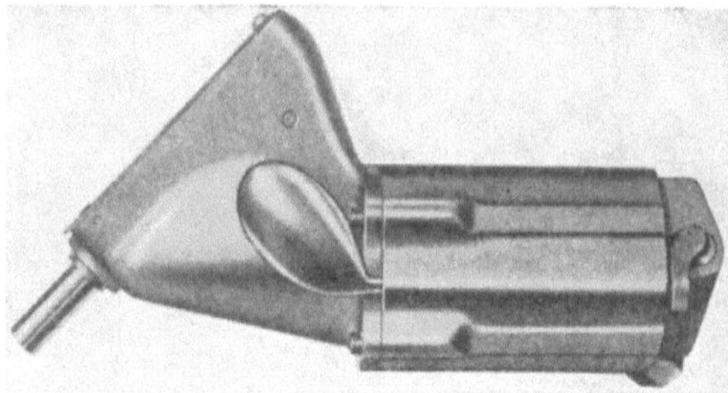

Fig. 224. A hammer designed for work in awkward places. (Courtesy The Cleveland Pneumatic Tool Company)

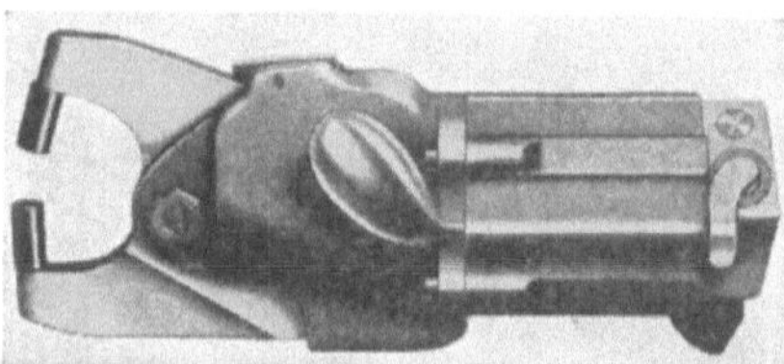

Fig. 225. An alligator squeeze riveter. (Courtesy The Cleveland Pneumatic Tool Company)

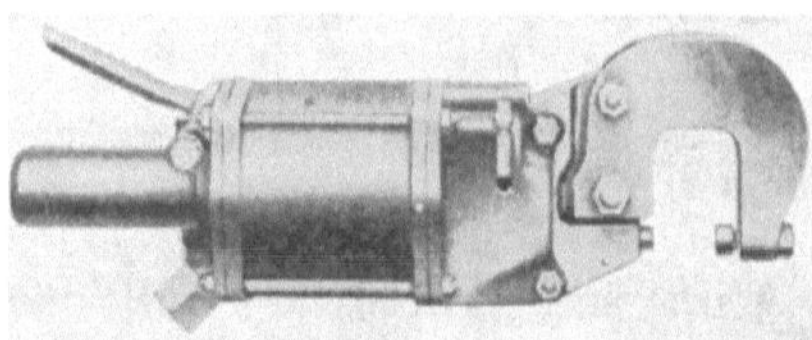

Fig. 226. A light-yoke squeeze riveter. (Courtesy The Cleveland Pneumatic Tool Company)

Fig. 227. A light-yoke squeeze riveter with handle. (Courtesy The Cleveland Pneumatic Tool Company)

RIVETING

Before driving rivets, the sheet-metal worker should check to be sure that there are no errors in part sizes and to see that all parts fit properly in place. The part should fit in place without forcing. Be sure there are no chips, dirt, or grit between the parts being riveted together. Examine the material to see that all burrs have been removed and that there are

Fig. 228. Operating a pneumatic squeeze riveter. (Courtesy The Cleveland Pneumatic Tool Company)

no scratches or other injury to the part. Check to see that priming coats, alclad coats, paint, etc., are in good condition before riveting.

Before driving a rivet, the rivet hammer should be adjusted, and it should be thoroughly cleaned before attaching the air hose. To adjust the gun, place the rivet set in the gun and see that the set retainer is in place. Press the rivet set against a block of wood laid on the top of a bench and depress the trigger for about one second. The set should be pressed against the wood with about the same pressure used when setting a rivet. The beginner will soon learn the proper adjustment and feel of the rivet gun by this method. Never pull the trigger unless the rivet set is against a block of wood or a rivet is being driven. If the trigger is pulled without a rivet set, damage to the gun may result.

Usually riveting is done by a team, the riveter and the bucker. The bucker holds the bucking bar against the shank end of the rivet, while the riveter "shoots" the rivet. The bucker is just as important as the riveter. It is his responsibility to see that the head is properly formed.

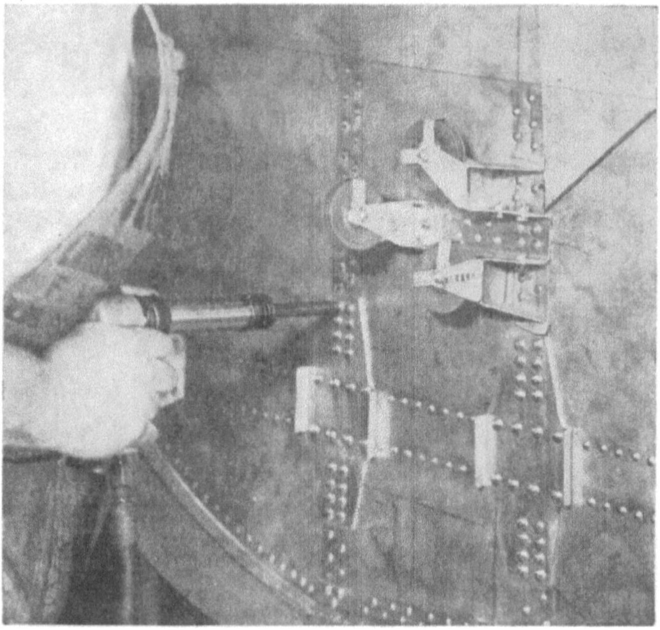

Fig. 229. Operating a one-shot riveting gun. (Courtesy The Cleveland Pneumatic Tool Company)

The bucking bar should never be removed from a rivet until the riveter stops shooting. It is often impossible to see what the other fellow is doing, and a set of signals is necessary between the bucker and the riveter. A light tap or slight pressure on the rivet tells the riveter that the bucker is ready. When the riveter feels that the rivet is set, he stops the gun. One tap by the bucker indicates that the rivet needs more setting; two taps that the rivet is properly set; three taps, the rivet is faulty and should be removed. Other signals may be arranged between the partners, but these signals are generally used.

The bucking bar must be held firmly against the rivet whenever the rivet hammer is in action. Rivets should be formed as rapidly as possible. Too many light blows may work-harden the rivet, causing cracks.

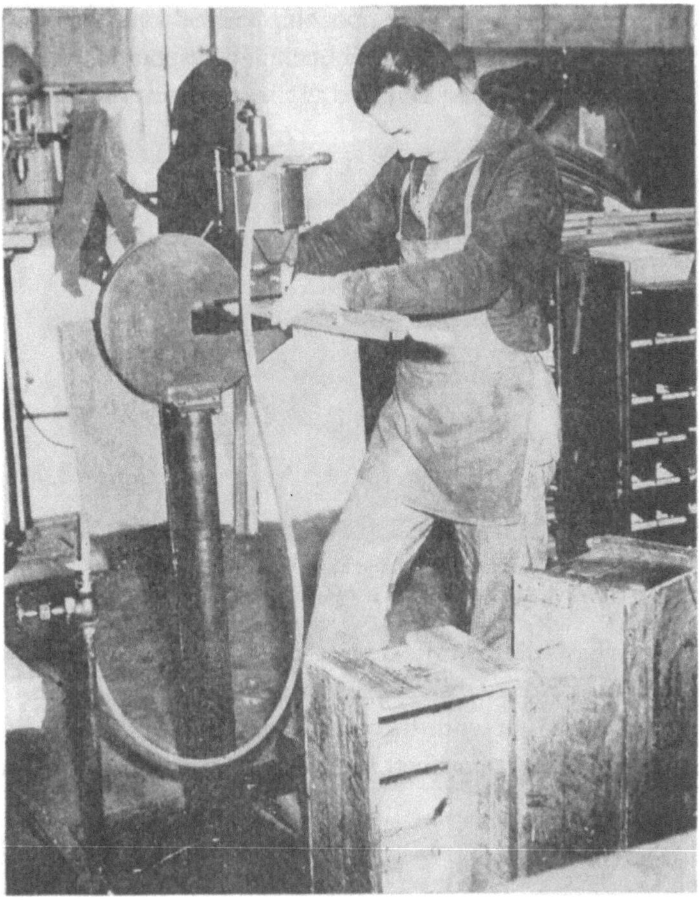

Fig. 230. Operating a yoke squeeze riveter mounted on a stand. (Courtesy The Cleveland Pneumatic Tool Company)

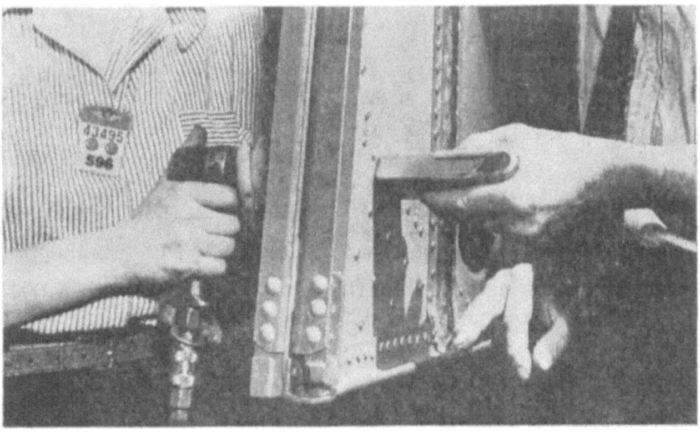

Fig. 231. A riveting team. Note the rubber pad held against the bucking bar to reduce vibration. (Courtesy Douglas Aircraft Company, Inc.)

The rivet set should, whenever possible, be held parallel to the axis of the rivet shank. The face of the bucking bar should be parallel to the surface of the part being riveted. It is sometimes necessary to use

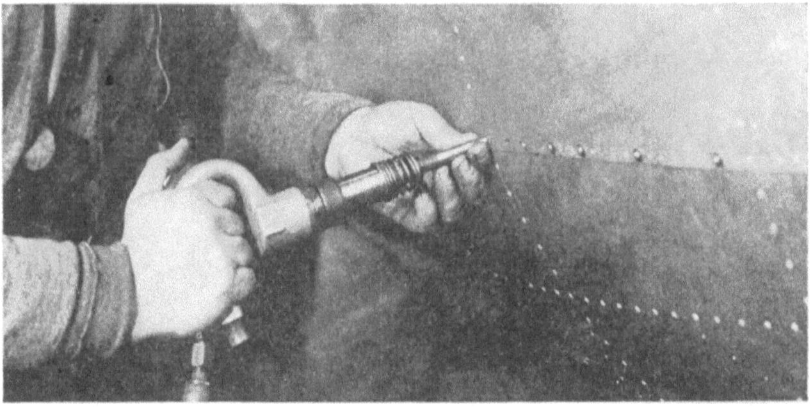

Fig. 232. Riveting an airplane fuselage skin. The bucker is inside the fuselage. (Courtesy The Cleveland Pneumatic Tool Company)

special, offset rivet sets in order to get into corners or out-of-the-way places. When an offset tool is used, as much of the pressure as possible should be applied in line with the rivet shank. If it is impossible to hold a bucking tool in line with a rivet, a special bucking tool with a slanting

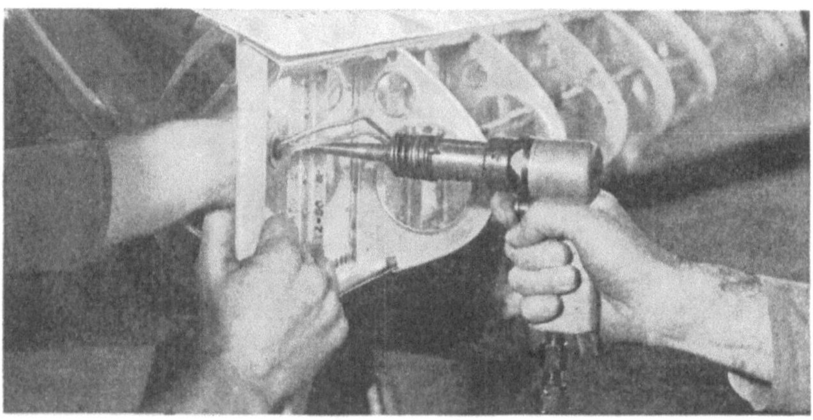

Fig. 233. Riveting and bucking sometimes have to be done in difficult positions. (Courtesy The Cleveland Pneumatic Tool Company)

face should be used. The weight of the rivet gun and the size of the set depend upon the size of the rivet to be driven. Large rivets require harder blows than do small rivets. The riveter can often develop the ability to determine when the rivet is properly set by the sound as well as the feel. As the rivet is upset, there is a slight change in the sound produced by the hammer against the rivet head.

A bucking bar should be as heavy as is consistent with the work. A light bar does not absorb the vibrations of the rivet hammer. It is sometimes necessary to buck a rivet which cannot be seen. Bucking bars are of many different shapes, depending upon the position of the rivet

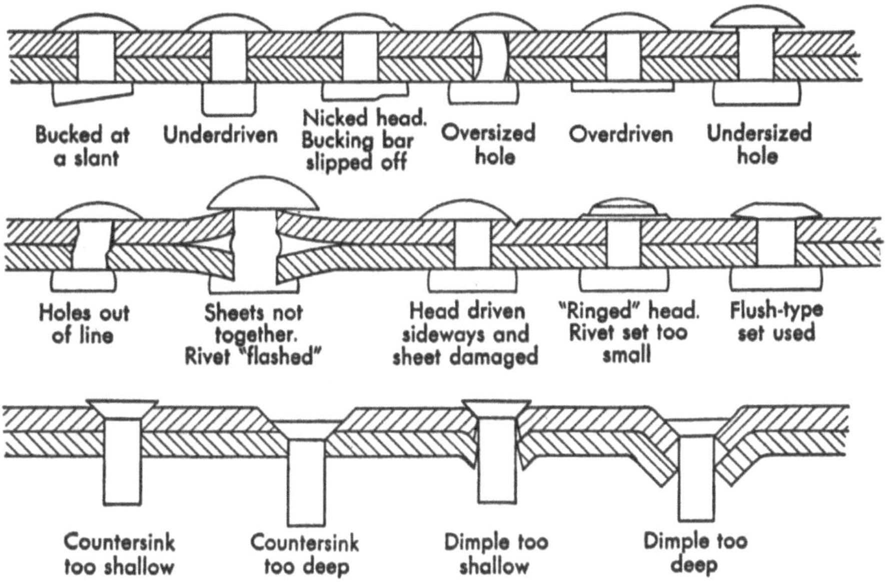

Fig. 234. Common defects in riveting.

in the assembly. Removing the bucking bar while the hammer is still in action may cause the rivet set to be driven through the sheet. The hammer operator and the bucker are both responsible for the job.

Rivets to be removed should be circled with a grease pencil. All cracked or damaged rivets must be removed. A deformed or defective rivet forms a weak spot in a riveted assembly. Some of the common defects found in rivets which are cause for removal are factory heads that are ringed, scored, scratched, or cracked; or the formed head may be too flat, formed crookedly, scored, scratched, or defaced. Rivets that have not been formed flat enough may have this condition corrected by additional driving. Sheets that have been riveted together with chips and dirt between them must have the rivets removed and the space properly cleared of foreign material. The riveter should watch carefully for buckles while riveting and, if buckles have a tendency to form, should call the condition to the attention of the supervisor.

Fig. 235. An instructor correcting the position of a rivet gun operated by a student. The instructor is holding the bucking bar. (Courtesy Douglas Aircraft Company, Inc.)

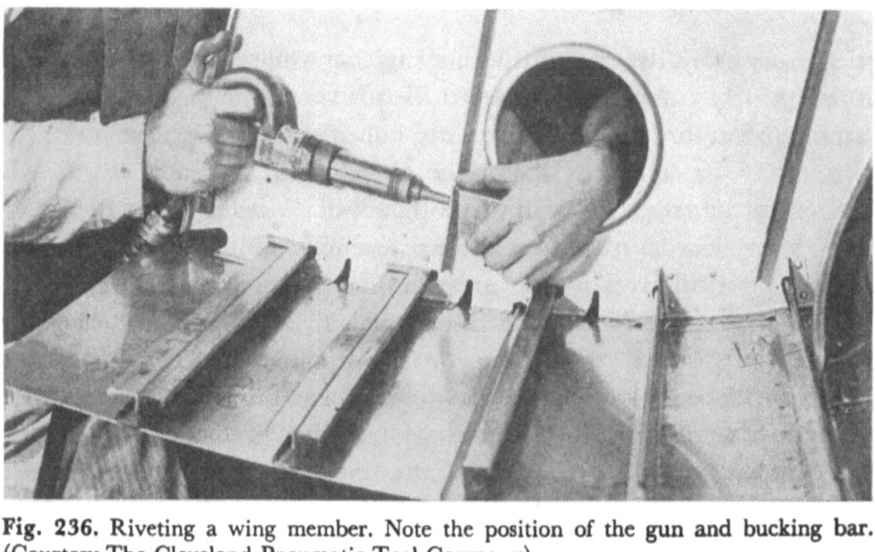

Fig. 236. Riveting a wing member. Note the position of the gun and bucking bar. (Courtesy The Cleveland Pneumatic Tool Company)

Rivets are usually removed by drilling through the head and driving out the rivet with a punch. A drill the exact size of the rivet should be used and it should be carefully centered on the rivet head. Do not drill past the surface of the sheet, as shown in Figure 237. The head is then either punched or sheared off with a cold chisel having rounded corners, and the rivet shank driven out of the hole with a suitable punch. Do not try to knock the head off a rivet with the rivet hammer. The driving of a rivet usually expands the rivet hole. If it is necessary to replace the rivet with one of a larger size, this should be called to the attention of the supervisor. When drilling out rivets, it is usually necessary to center-punch the rivet head to prevent the drill from wandering off the head. The rivet head will usually come off on the drill as soon as the drill has reached the required depth, if the drill has been properly located. If light blows do not drive out the rivet, the sheet must be backed up to prevent damage. More rivets have to be replaced because of poor bucking than because of poor driving.

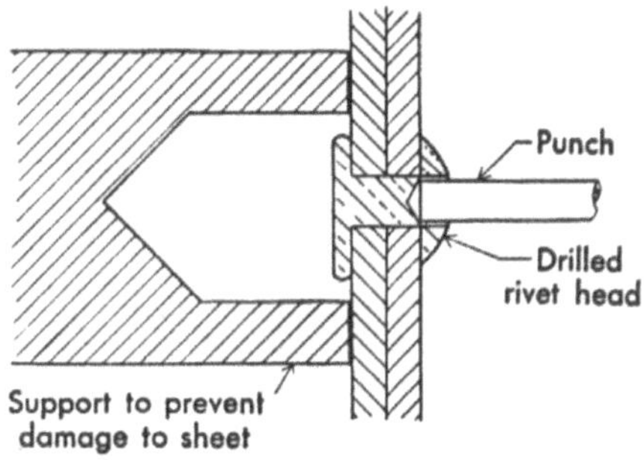

Fig. 237. A rivet drilled for removal.

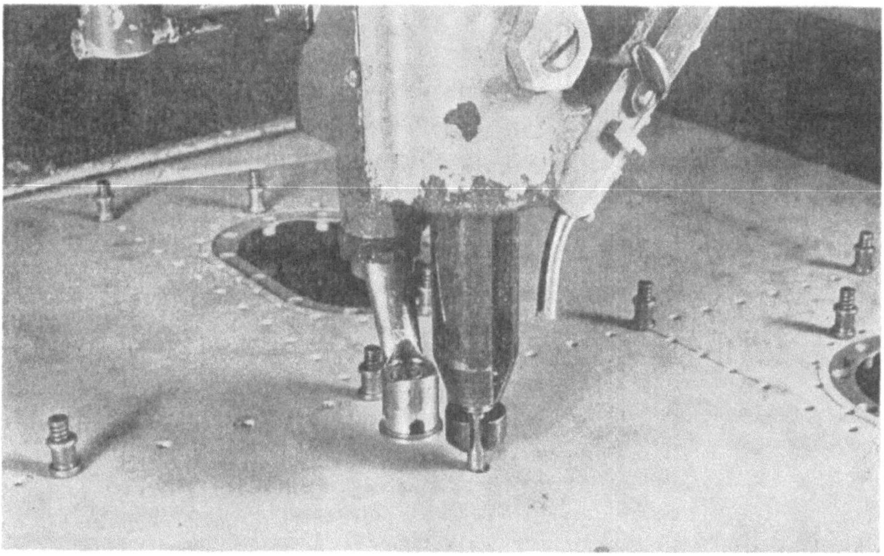

Fig. 238. Operating a riveting machine. Note the dimpled holes. (Courtesy Douglas Aircraft Company, Inc.)

Countersinking for flush-type rivets is done by a countersink drill. These drills must have the exact angle required by the rivet. The countersinks should be equipped with a stop which prevents the countersink

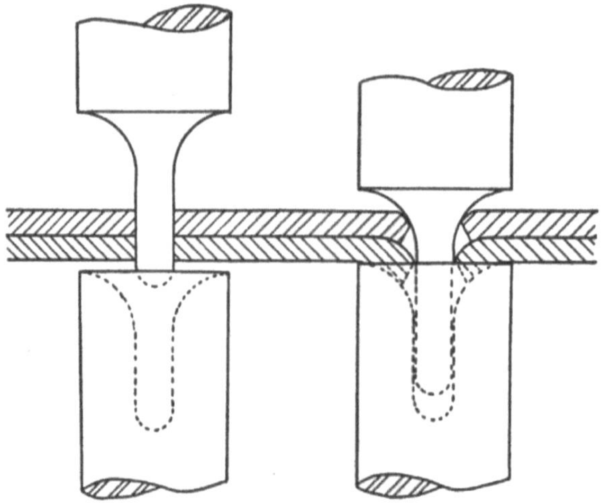

Fig. 239. Dimpling dies.

becoming too deep. Usually countersinking is not done on sheet having a thickness of less than 0.050 in. On sheet of less than the above thickness, the skin is usually dimpled, as shown in Figure 238. This dimpling may be done either with special dimpling dies, as shown in Figure 239,

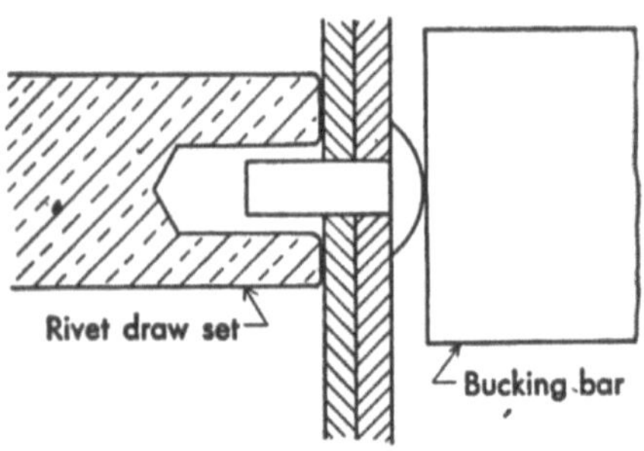

Fig. 240. A rivet draw set.

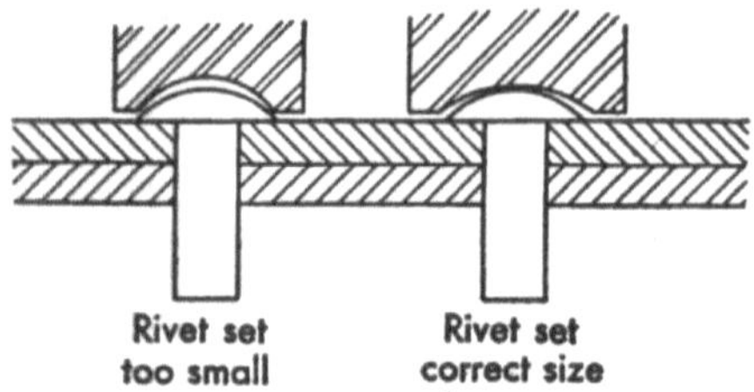

Fig. 241. Incorrect and correct sizes of rivet sets.

or, on very thin sheet, the rivet itself may be used to form the dimple. Dimpling is usually done with a machine.

When rivets do not slip freely into place or the sheets do not fit closely together, it may be necessary to use a draw set to seat the rivet properly. A draw set is made of wood, fiber, soft metal, or some material which will not injure the sheet. A hole is drilled in the end the size of the rivet, but considerably deeper than the length of the rivet shank which protrudes through the sheet. In drawing a rivet, the head of the rivet should be backed up to hold the rivet and sheet firmly in place. The draw set is placed over the rivet on the opposite side of the sheet and tapped lightly

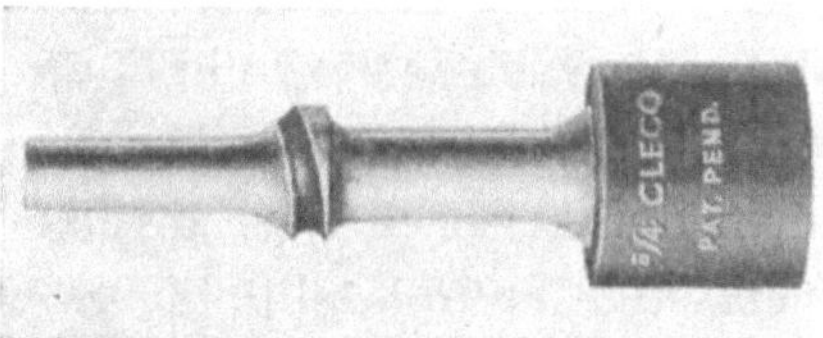

Fig. 242. A swivel flush rivet set. Note the groove for the set retainer spring. (Courtesy The Cleveland Pneumatic Tool Company)

Fig. 243. Parts for the swivel flush rivet set. (Courtesy The Cleveland Pneumatic Tool Company)

with a hammer or mallet. Sufficient force to damage the sheet should not be applied.

In driving round-head rivets, the curve of the set should be slightly greater than the curve of the rivet head. Sets for flush-type rivets should be flat or slightly crowned, but the edges should be rounded and the surface highly polished to prevent injury to the sheet. In driving counter-

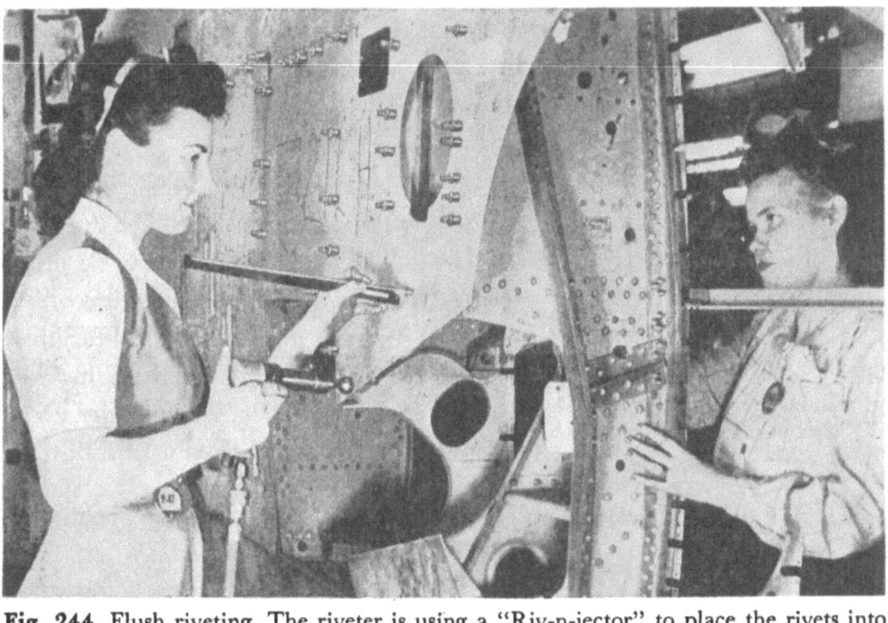

Fig. 244. Flush riveting. The riveter is using a "Riv-n-jector" to place the rivets into the holes. Note the Cleco clamps holding the sheets in place. (Courtesy The Cleveland Pneumatic Tool Company)

sink rivets on curved surfaces, it may be necessary to rock the hammer slightly from side to side to make the head match the curve of the material. The edges are slightly rounded to prevent damage to the sheet. This type of set must be held exactly parallel to the surface of the material being riveted. In order to do this, a swivel type of set, such as shown in Figure 242, is used. The cylindrical rubber part holds the two parts of the set together and prevents the set from slipping over the surface being riveted when the gun is held at an angle. Figure 244 shows flush riveting. The Riv-n-jector is a tool in which a number of rivets are loaded. At the end of the tube is a special chuck which allows the ejection of one rivet at a time.

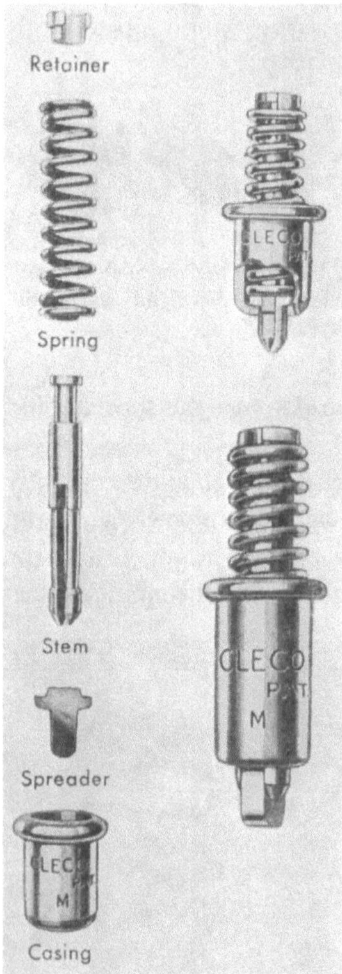

Fig. 245. A Cleco sheet holder and its various parts. (Courtesy The Cleveland Pneumatic Tool Company)

It is often necessary to use some means to hold the sheets closely together, or some such tool as the Cleco sheet holders. These holders are inserted by means of a special pair of pliers and hold the sheet firmly in place while riveting. Figure 245 shows the Cleco sheet holder and its various parts.

It is good practice to cover the rivet set with masking tape to protect the rivet head and the sheet from injury. A number of rivets are often placed in position and covered with a strip of masking tape which prevents them from jumping out due to the vibration of driving the rivet. Most rivets may be driven without removing the masking tape.

When the holes through which the rivet must pass in two sheets do not exactly match, it may be necessary to use a tapered awl to line up the material. Care should be taken not to apply enough force to deform the sheets.

Blind riveting is done where it is impossible to buck a rivet. Blind rivets are inserted and set from one side of the part being riveted. Most of these rivets are constructed of two or more parts and require a special

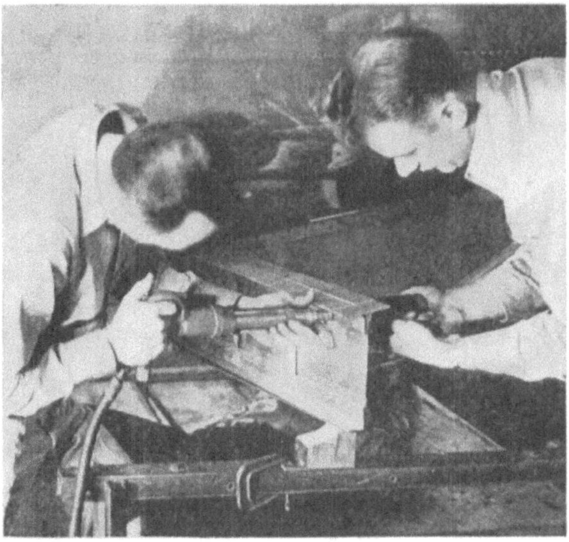

Fig. 246. Riveting an airplane wing spar. (Courtesy The Cleveland Pneumatic Tool Company)

tool for setting. Blind rivets are not driven. One type of blind rivet has a small explosive charge in the hollow shank. This charge is exploded by a hot, silver-tipped, rivet set applied to the head of the rivet. Blind

Fig. 247. Riveting a structural member. (Courtesy The Cleveland Pneumatic Tool Company)

rivets are only used when it is impossible to drive a rivet in the usual manner. Blind rivets are commonly used in making repairs such as a patch on a wing or control surface.

Fig. 248. A heavy-riveting machine. (Courtesy Douglas Aircraft Company, Inc.)

Metal screws which are self-threading are also used in places where it is impossible to drive a rivet properly. The hole is drilled somewhat smaller than the screw. The screw is driven into place, cutting its own thread.

Heavy riveting on structural parts is often done by heavy-riveting machines such as shown in Figure 248. This is a yoke-type riveter operated by compressed air. It is portable and may be moved by overhead trolleys. The jaws of the machine may be adjusted to various positions by rotating the entire machine on the tracks in back of the yoke and at the top near the suspension hook. A smaller squeeze riveter to be used for riveting structural parts is shown in Figure 249. This riveting hammer is mounted on an overhead beam and sets the rivet with a

Fig. 249. A heavy squeeze-riveting machine. (Courtesy Aluminum Company of America)

single stroke of the plunger. The machine in Figure 250 is operated by an electric motor. The blows are struck by the hammer-like arrangement at the top of the machine. This hammer is mounted on a piece of wood which resembles the handle of an ordinary hammer. The manufacturer states that the elasticity of the wood has the same effect as riveting by hand, using a wooden-handled hammer. This machine is used for cold riveting and will handle both ferrous and non-ferrous rivets. An advantage of this machine is that the rivet set rotates continuously while the machine is in operation, decreasing the danger of cracked rivet heads.

Problem 1. Using two 6-inch squares of 2S aluminum alloy, 0.040 in. thick, lay out and drill holes as shown in Figure 251, using 2S–O rivets.

Fig. 250. A motor-driven riveting machine. (Courtesy High Speed Hammer Co.)

Insert the rivets through the holes and clamp the sheets to a steel plate with the factory heads against the plate. Using the flat face of a ball-peen machinists' hammer, form the proper head on the shank ends of the rivets.

Problem 2. Lay out and drill the proper rivet holes in 6-in. squares of 17S–T aluminum alloy having a thickness of 0.063 in. Rivet the sheets together in the same manner as in Problem 1. Use 17S–T rivets which have been heat-treated. These rivets must be driven before they have had time to harden. Drive a few rivets in the 17S–T condition that have

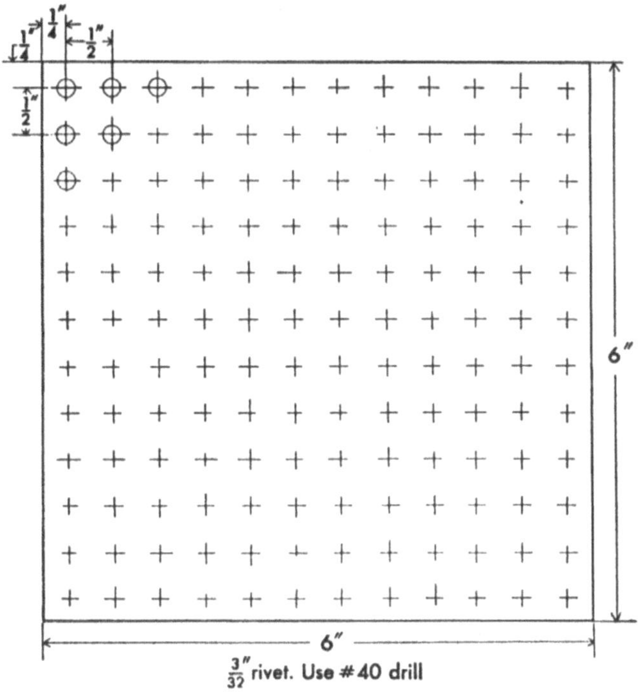

Fig. 251. A sheet with rivet holes laid out for drilling.

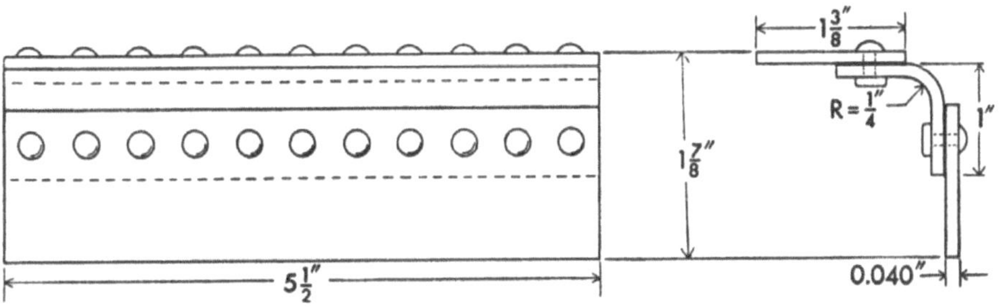

Fig. 252. Two edges of a flange riveted to sheets.

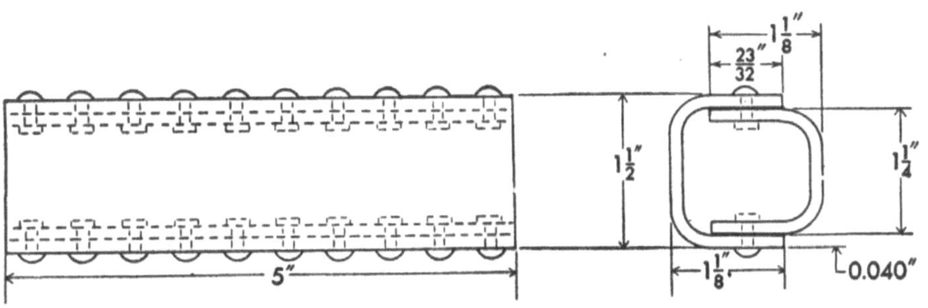

Fig. 253. Two channels riveted together.

had time to harden. Note the difficulty with which these rivets are driven.

Problem 3. Repeat Problem 1, using a hand set to form the head on the shank end of the rivets.

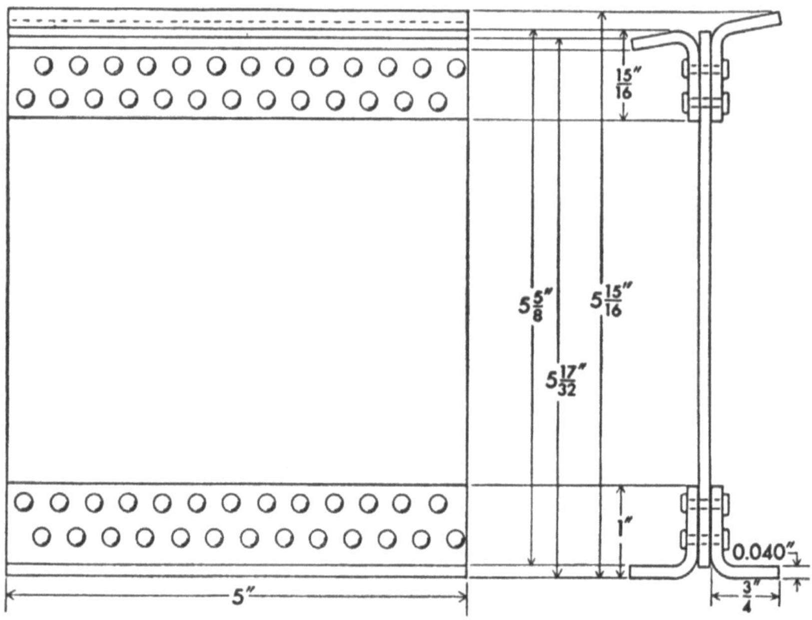

Fig. 254. A front spar section.

Problem 4. Repeat Problem 2, clamping the sheets in a vertical position and set the rivets, using a rivet gun. The riveter may do his own bucking or have another person hold the bucking bar.

Problem 5. Form a flange having the dimensions shown in Figure 252 of 17S–O aluminum alloy, and rivet both edges of the flange to strips of 17S–O. All material should be 0.040 in. in thickness. Use 17S–T rivets. These rivets should be driven with a slow-hitting rivet gun.

Problem 6. Using 17S–O having a thickness of 0.040 in., rivet two channels together as shown in Figure 253. This operation requires a square bucking bar which is slipped through the opening formed between the channels. The rivets should be 17S–T. Use a fast-hitting rivet gun.

Problem 7. Form a front spar section of 24S–O having a thickness of 0.040 in. and the dimensions shown in Figure 254. The rivets should be 24S–T and driven before they have had time to harden. Use a one-shot rivet gun and flat-head rivets.

Problem 8. Form a rear spar section having the dimensions shown in Figure 255 of 24S–T. The rivets should be 24S–T of the same style and driven in the same manner as those used in Problem 7.

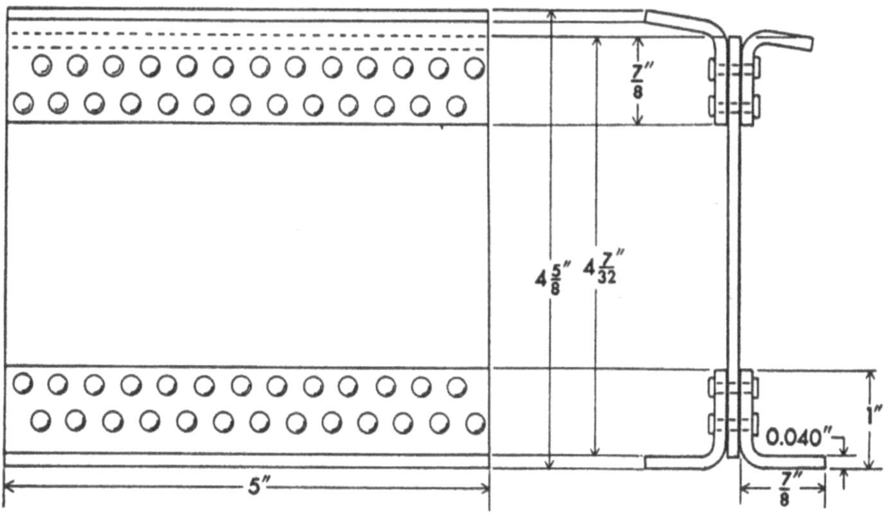

Fig. 255. A rear spar section.

Problem 9. Using the rib parts formed in Chapter XVI, Figure 195, rivet the nose, web, and trailing end to the spar sections formed in Problems 7 and 8, as shown in Figure 256.

Problem 10. To represent the skin of a wing, rivet strips of 24S–T, 2 in. wide, 0.015 in. thick, to the rib assembled on the spar sections in Problem 9, using flush-head, 24S–T rivets. After the strips are put in place and the proper rivet holes drilled, the holes should be dimpled before setting the rivets. The strips around the leading edge should overlap the strips extending from the front spar to the trailing edge.

Experience should be had in flush riveting before attempting to do this kind of riveting on actual construction or repairs. (1) Practice flush riveting on thin sheet where the rivet forms the dimple as it is driven. The proper dies and sets should always be used. (2) On heavier sheet preform the dimples before setting the rivets. (3) Practice flush riveting of a dimpled sheet to a heavy sheet in which the rivet holes have been countersunk. (4) Rivet thin sheet to heavy sheet in which the holes have been countersunk, dimpling the thin sheet while setting the rivet. Flush riveting is more difficult than regular riveting as it requires a higher degree of skill not to have "high" or "low" rivet.

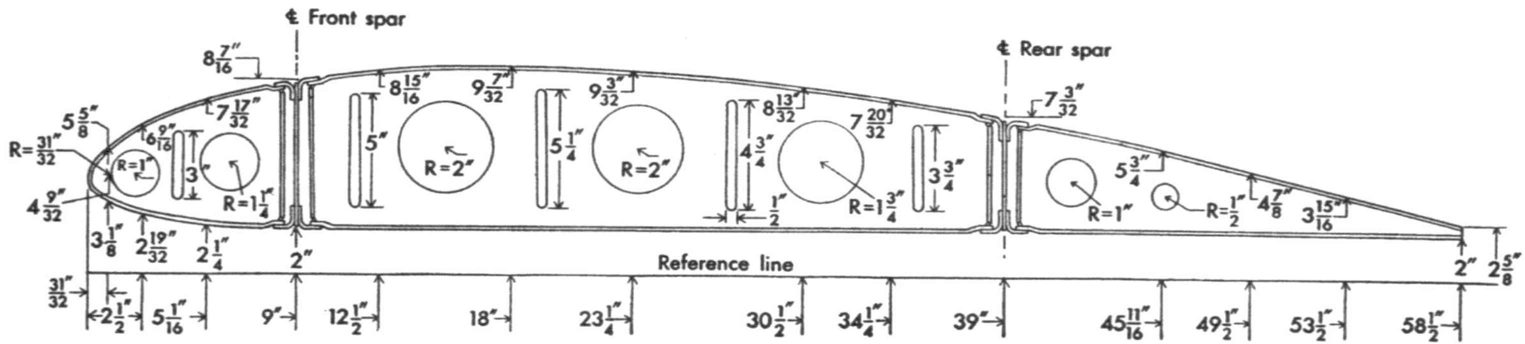

Fig. 256. Assembled rib parts formed in Chapter XVI.

XIX FORMING MAGNESIUM ALLOYS

Most magnesium parts may be formed in one operation, thus avoiding repeated annealing and redrawing. In forming magnesium at elevated temperatures, the deep-drawn parts do not require additional dies. This constitutes a saving in both time and equipment. When formed at approximately 600° F., no allowance need be made for springback. If worked at lower temperatures, springback allowance must be made. For example, Dowmetal J–1H sheet, when bent around a radius four times the thickness of the metal at 400° F., must have a springback allowance of approximately 3°. Table VIII shows the most satisfactory forming temperature ranges for magnesium alloys in both the hard-rolled and annealed condition.

TABLE VIII. RADII * FOR 90° BENDS IN MAGNESIUM ALLOY SHEET

DOWMETAL ALLOY	CONDITION	MINIMUM RADII AT 70° F.	TYPICAL RADII AT VARIOUS TEMPERATURES			
			70° F.	*300° F.*	*400° F.*	*600° F.*
FS–1a	Annealed	5T**	4½T	—	1.2T	—
FS–1h	Hard rolled	10T	8T	4½–6T	—	—
J–1a	Annealed	9T	7T	—	4–6T	2–3T
J–1h	Hard rolled	17T	13½T	—	4½–6½T	—
Ma	Annealed	7T	5½T	—	3–4T	1–2T
Mh	Hard rolled	12T	9T	—	6–7T	—

* Radii formed in specimens 6 in. wide in square rubber retaining dies.
** T = thickness of sheet.

Magnesium alloy sheet may be sheared, punched, and blanked like sheet of other metals. When magnesium sheet thicker than approximately 0.0464 in. is sheared or punched, the edges show a characteristic, rough, flaky fracture. This rough edge may be lessened by proper clearance on shears and punches. Maximum clearance of from 3 to 5 per cent of the sheet thickness is recommended. The sheared edge may be smoothed by a double shearing operation. This operation consists of removing approximately 1/16 in. of the material by a second shearing. Hot shearing of heavy sheets and plate improves the condition of the sheared edges. Annealed sheet may be heated to approximately 600° F. for shearing. Other alloys may be heated as follows: FS–1H, 275° F.; J1–H, 400° F.; and MH, 400° F. These alloys are hard rolled. Due to the expansion of the metal when hot, a small allowance must be made when shearing. Magnesium thicker than ½ in. should be sawed rather

Fig. 257. Airplane wheel fairings formed of magnesium alloy. (Courtesy The Dow Chemical Company)

than sheared. The oxy-acetylene cutting blowpipe cannot be used for cutting magnesium alloys.

Magnesium alloy sheets may be formed by either hydraulic or mechanical presses. Hydraulic presses are usually preferred as their controllable speed is an advantage. Slow drawing speeds are desirable when forming magnesium. These alloys may be formed on the ordinary brakes or bending rolls, but the press-type brake is preferred, as the leaf-type brake tends to concentrate the bending at the clamped edge, causing fractures. The dies on the press-type brake are more easily heated than on the leaf-type. Drawing dies for magnesium sheet are similar to those for other metals with the exception that a provision for heating the parts in contact with the metal is necessary. This heating is usually done by electric heaters held in contact with the dies. The dies may be made of mild steel, tool steel, nitrided, or chromium-plated steels. The dies should be highly polished and well lubricated. The coefficient of expansion of magnesium alloys is approximately three times that of steel. For this reason dies of steel for hot forming must be slightly oversized

Fig. 258. A press used to form wheel fairings of magnesium alloy. (Courtesy The Dow Chemical Company)

in order that the finished part will have the proper dimensions. When dies are made of aluminum alloys, an allowance must also be made as the coefficient of expansion of aluminum is less than that of magnesium.

Forming dies are generally heated with gas burners which are automatically controlled. Lubrication is important in the forming of magnesium alloys to prevent scoring or marring of the drawn part. The lubricant should have the following qualities: it should be easy to apply evenly and smoothly; have good lubricating qualities; have good film strength; be able to withstand temperatures from 600 to 700° F.; be as chemically inert as possible; leave no heavy or burned-on residue; and have no chemical or physical reaction with the alloy.

The surface of the dies and the material to be formed should be thoroughly lubricated before forming is started. Colloidal graphite suspended in a suitable carrier is an ideal lubricant for magnesium sheet and can be applied by spraying. A mixture of 20 per cent graphite in

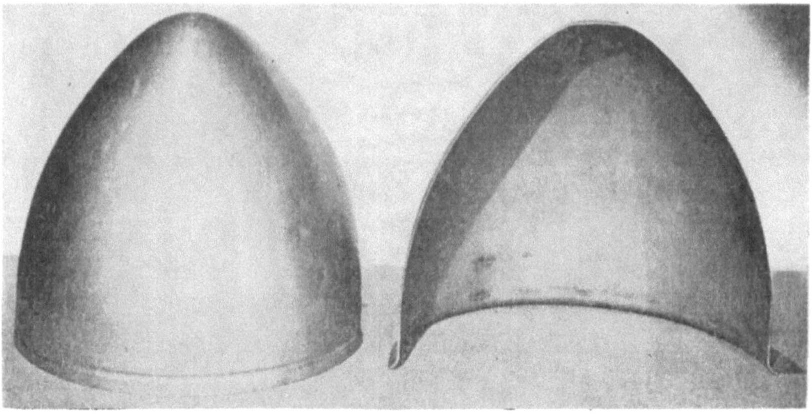

Fig. 259. Airplane nose spinners formed of magnesium alloy by deep drawing. (Courtesy The Dow Chemical Company)

tallow is recommended for use on the dies and is applied by buffing to the die surface. This lubricant may be used on material to be formed, but it is not as suitable as the colloidal graphite suspension. If colloidal graphite is used, the sheet stock should not have chrome-pickled surfaces as it is difficult to remove the graphite. Sheet should be ordered in the oiled rather than the chrome-pickled condition when graphite is to be used as a lubricant. If graphite must be used on a chrome-pickled sheet, the pickled surface should be removed before forming by dipping from 1 to 15 min. in the following solution at room temperature: chromic acid, 1.5 lb.; water to make 1 gal. After dipping in this solution the material should be thoroughly washed in clear water and dried. After forming operations, the part should be thoroughly cleaned of all lubricants by means of a vapor degreaser or alkaline cleaner. This is followed by cleaning in the same chromic acid solution used to remove the chrome-pickled surface. The chromic acid cleaning solution will not remove graphite which requires a special cleaner. Parts may have the graphite removed by use of the following solution: chromic acid, 1½ lb.; calcium nitrate or magnesium nitrate, 4 oz.; water to make 1 gal. This solution is used at room temperature and the parts are dipped into it for 2 to 5 min. The parts are then washed in cold water, then dipped in hot water to assist in drying. This solution will also remove oil-type lubricants. Both cleaning solutions must be kept in pure aluminum tanks.

Magnesium sheet which is to be hot formed should be preheated to the working temperature. This heating may be done in gas or electric furnaces or by contact heaters such as hot plates. Heated baths may

Fig. 260. A magnesium alloy oil pan formed on a hydraulic press. (Courtesy The Dow Chemical Company)

also be used. Preheating makes sure that the proper drawing temperature is obtained and prevents distortion which might occur if placed cold in contact with the hot dies. The preheated material assists in keeping the dies at an even temperature. Rubber may be used in hydraulic presses to form magnesium sheet in a manner similar to that used in forming aluminum alloys. Special rubber compounds or synthetic rubber products are used, and the sheet may be formed up to temperatures of approximately 450° F. Cornstarch may be used to prevent the rubber sticking to the parts at these temperatures.

By the use of heated dies, magnesium alloys in the sheet form may be bent in the same manner as aluminum alloys. Table IX gives the radii for 90° bends in magnesium alloy sheet at various temperatures.

TABLE IX. FORMING TEMPERATURES FOR MAGNESIUM ALLOYS

DOWMETAL ALLOY	CONDITION	WORKING TEMPERATURE RANGE
		Degrees F.
FS–1a	Annealed	400–500
FS–1h	Hard rolled	275 or 300 max.*
J–1a	Annealed	550–650
J–1h	Hard rolled	400 max.
Ma	Annealed	550–650
Mh	Hard rolled	400 max.

* 300° F. may be used if metal is at temperature 15 min. or less.

Magnesium alloys can be hand formed in the same manner as the aluminum alloys. The material must be heated and should be hammered with a leather mallet. In hand forming, it is suggested that metal form blocks and backing blocks be used. The material to be formed is clamped between the blocks, and both the blocks and the metal heated to the desired temperature. The blocks will maintain the heat of the part being

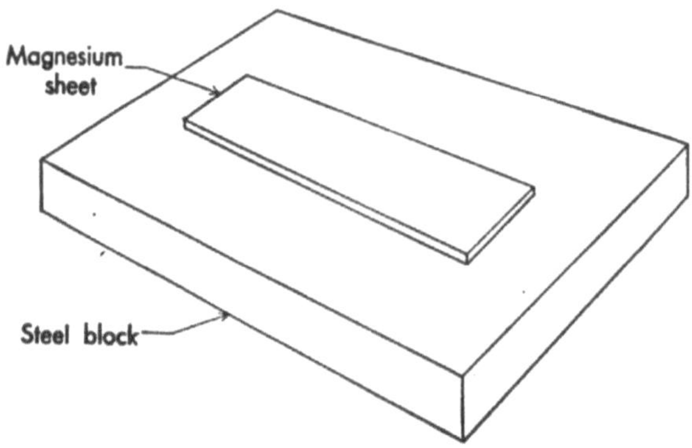

Fig. 261. Stretching magnesium sheet on a metal plate.

worked for a considerable period of time. A leather mallet is recommended for handworking magnesium although, where necessary, a mallet made of soft wood may be used.

Problem 1. Obtain a piece of Dowmetal sheet approximately 1 in. wide and 6 in. long, having a thickness of about 0.063 in. Place the material on a smooth steel plate, as shown in Figure 261. With an ordinary blowtorch, heat both the material and the plate to approximately 400° F. Magnesium alloys show no color changes when heated, and steel does not begin to show a dull red until heated to approximately 900° F. Since

400° F. (less than half the temperature at which steel shows red) is only twice the boiling point of water, a drop of water on the metal block will sizzle violently at approximately 350° F. Care should be taken not to overheat the material. With a steel finishing hammer, stretch the magnesium sheet in the same manner as described for the stretching of aluminum alloys in Chapter XV.

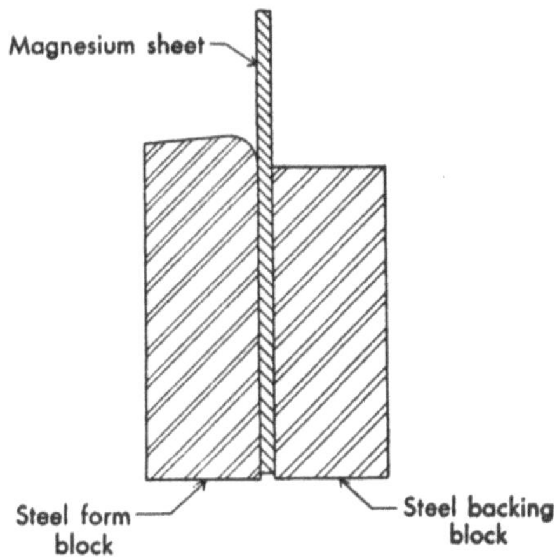

Fig. 262. Forming a flange in magnesium alloy sheet.

Problem 2. Obtain a piece of Dowmetal sheet approximately 0.040 in. in thickness, 4 in. wide, and 6 in. long. Using a metal form block as shown in Figure 262, clamp the material between the blocks, heating the blocks and material, as described in Problem 1. Using a leather mallet, form a 90° flange.

Problem 3. Using a curved form block and backing block of hardwood or similar material, form a curved flange having the dimensions shown in Figure 263 from Dowmetal sheet having a thickness of approximately 0.030 in. Note that the temperature to which the metal is to be heated is below the charring point of wood. Wood begins to char at approximately 800° F. The material being

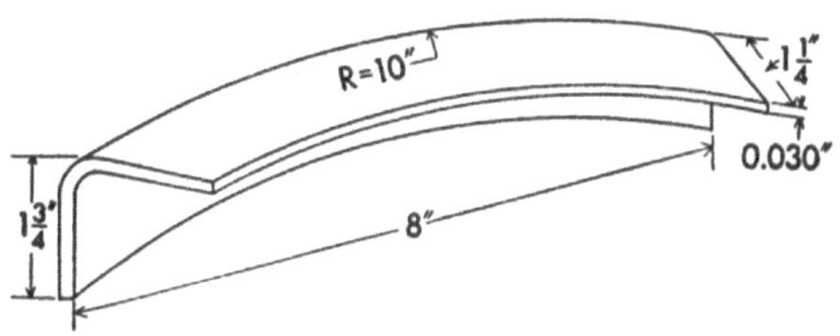

Fig. 263. Forming a curved flange by shrinking one side of the angle.

formed may be heated to the proper temperature by passing the flame from the blowtorch back and forth over the metal without charring the wood. If the metal cools and becomes difficult to work, it may be reheated. It is better to reheat slightly below the maximum working temperature than to overheat.

XX FORMING CORROSION-RESISTANT ALLOYS

The most important of the corrosion-resistant alloys used in aircraft sheet-metal work is stainless steel. The alloy, 18–8, is perhaps the most commonly used of the stainless steels. In working with stainless-steel sheets the same general methods for forming are used as for other alloys. Stainless steels, however, are much more difficult to work, due to their toughness and hardness.

Hand forming of stainless steel is not done to any great extent. When these alloys must be formed by hand, it is usually necessary to form

Fig. 264. Pouring molten stainless steel from an electric furnace. (Office of War Information photograph by Al Palmer from a negative now in the Library of Congress)

them at a high temperature and then reheat-treat to restore the original qualities of the material. Almost all steels must be reheat-treated when they have been heated to a high temperature for any reason.

When forming by machinery the chromium nickel alloys lend themselves well to forming and drawing operations. They are ductile and have a high tensile strength. These alloys are somewhat stiffer than ordinary steels, and most of them harden much more rapidly when cold-worked.

Alloyed sheets which are to receive severe working should be specified to the manufacturer as "for deep drawing." Sheets of stainless-steel alloys should be in the annealed or dead-soft condition. Alloys of steel do not flow as readily in forming as do the aluminum alloys, but by proper tooling and careful working satisfactory results may be obtained. When extremely deep draws are to be made, it may be necessary to complete the operation in several steps, annealing the material between each step. It has been recommended that comparatively slow press speeds be used. These alloys, due to their high elastic limit, have considerable springback. Where springback cannot be allowed for in the

Fig. 265. Stainless steel passing through the rolls in a white-hot condition. (Office of War Information photograph by Al Palmer from a negative now in the Library of Congress)

original forming operation, it is sometimes necessary to perform an additional operation after the final annealing to set the part in its required form.

Alloys of steel may be successfully riveted. Small rivets having a diameter of 3/16 in. or less may be driven cold. Due to the tendency to harden

Fig. 266. First forming operation on a stainless-steel fitting. (Courtesy The Glenn L. Martin Company)

when worked cold, it is recommended that the rivets be set either by a squeeze riveter or by as few blows as possible. A single blow is more desirable than a number of light blows. Larger-sized rivets should be driven while hot. The temperatures at which the rivets are heated should be accurately controlled. The proper temperature is approximately 2000° F. The rivet should be set before the temperature drops below 1600° F. The temperatures at which rivets of the stainless-steel alloys are driven vary with the different alloys and should be checked carefully with the manufacturer's specifications. Heating for too long a time tends to cause scale to form on the material. Rivets of the chrome nickel alloys should not be heated for more than approximately 10 min., and the straight chromium alloys should not be heated more than approximately 25 min. Rivet holes may be either drilled or punched. It is necessary to ream punched holes. Drilled holes are preferred as they cause less strain to the material.

Fig. 267. Final forming operation on the stainless-steel fitting shown in Fig. 266. (Courtesy The Glenn L. Martin Company)

It is somewhat difficult to cut the stainless-steel alloys on shears. These alloys usually cut about one-quarter of the distance through, and the rest of the material breaks. Shear blades should be sharp and carefully adjusted to prevent drawing the metal over the edge of the blades.

When excessive strains are not to be encountered, stainless-steel alloys may be soldered or brazed. These alloys lend themselves particularly well to spot welding, and this method of fastening is used wherever possible. In making repairs to stainless-steel structures, it is not usually possible to spot-weld. It is allowable to use rivets, bolts, or screws. When stainless-steel sheet is not available, repairs may be made by using chromium molybdenum sheet having equal strength.

Inconel, another of the corrosion-resistant alloys, composed of nickel, chromium, and iron, is used in the aircraft industry for exhaust collector

Fig. 268. A slab of stainless steel about to enter the rolls. (Office of War Information photograph by Al Palmer from a negative now in the Library of Congress)

Fig. 269. Stainless-steel sheet formed by rolling. (Office of War Information photograph by Al Palmer from a negative now in the Library of Congress)

rings, manifolds, and other parts subjected to excessive heat. This material is worked in much the same manner as the stainless steels. It should be worked in the annealed condition. Inconel cannot be hardened by heat treatment, but may be readily hardened by cold-working. To anneal inconel, it should be heated to approximately 1800° F. for about 15 min.

Fig. 270. Finished stainless-steel sheet. Note the high polish. (Office of War Information photograph by Al Palmer from a negative now in the Library of Congress)

The material is then quenched in cold water. When the part has been subjected to excessive cold-working, it should be annealed by the above method to relieve internal stresses. This material does not strain-harden as rapidly as stainless steel. Inconel bends more readily than the stainless steels and should bend through an angle of 180° with or against the grain without cracking. The radius of the bending angle should be equal to one-half the thickness of the material. Inconel may be welded by the oxy-acetylene, electric arc, spot, or seam process. Parts are usually joined by welding rather than by riveting.

The working characteristics of Monel metal are similar to those of inconel.

Copper is ductile when in the annealed condition and is easily worked. It becomes hardened through cold-working and must be re-annealed when worked excessively. Copper may be annealed by heating to a red heat, which is approximately 1000° F., and quenching in cold water.

Copper parts in direct contact with aluminum alloys should be cadmium plated. This does not apply to pressed fit bushings or threaded connections.

Sheet brass in the annealed condition may be formed readily. When

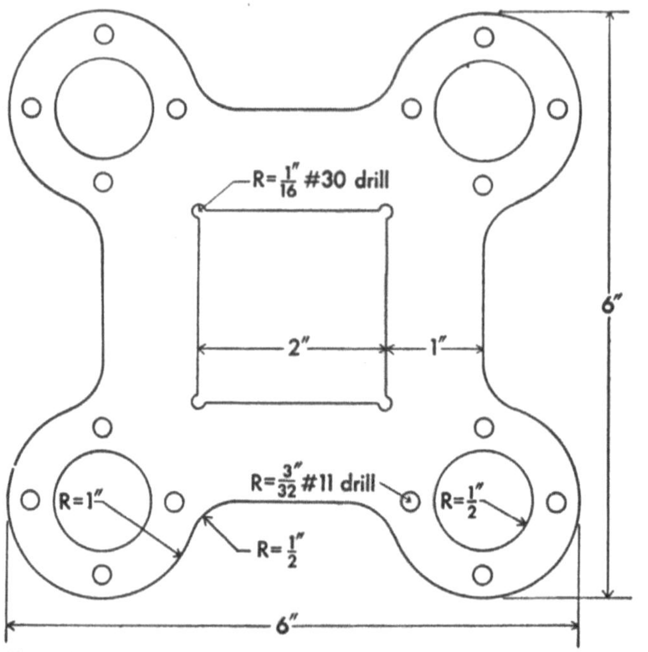

Fig. 271. A nickel steel fitting.

brass parts come in contact with aluminum alloys, they should be cadmium-plated. Copper and brass sheets may be joined by soft solder, hard solder, or brazing.

All corrosion-resistant alloys should be worked and treated in accordance with the manufacturers' specifications.

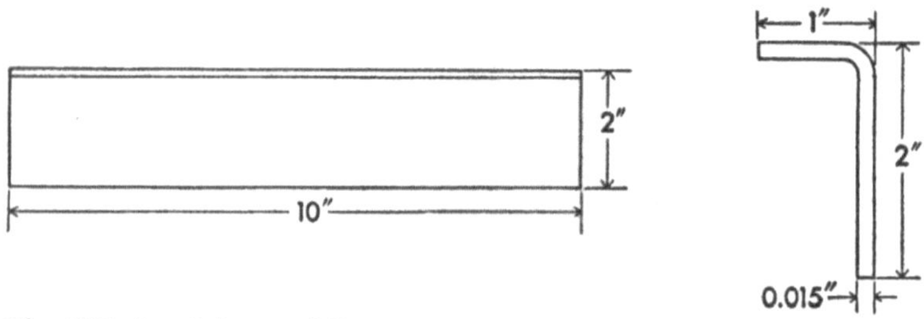

Fig. 272. A stainless-steel flange.

Problem 1. Obtain a piece of nickel steel about 6 in. square having a thickness of approximately 0.125 in. Form the fitting shown in Figure 271. Before starting to work on this material it should be heated to approximately 2000° F. and allowed to cool slowly. If properly heated and cooled, the material will be in the annealed condition. After forming, the material should again be heated to approximately 2000° F. and quenched in cold water.

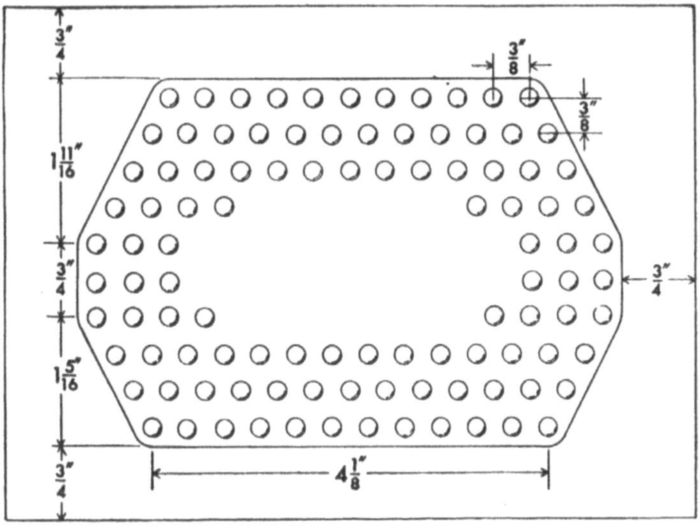

Fig. 273. A stainless-steel patch.

Problem 2. Obtain a piece of 18–8 stainless steel approximately 3 in. wide and 10 in. long having a thickness of 0.015 in. Form the flange shown in Figure 272. This material should be annealed before working.

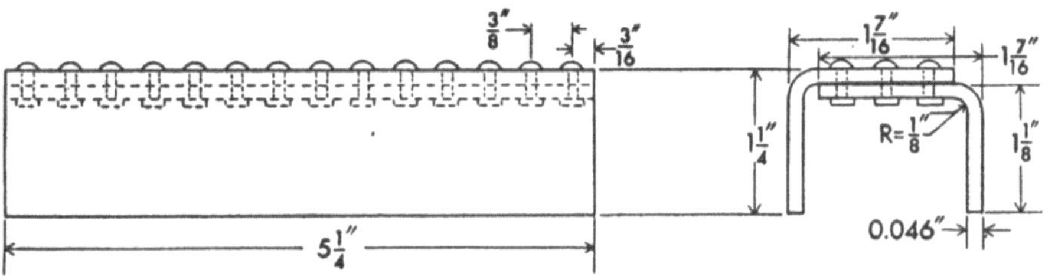

Fig. 274. Two flanges riveted together.

Problem 3. Using the thinnest 18–8 stainless-steel sheet available, prepare a patch as shown in Figure 273. The rivets used should be stainless steel and approximately 1⁄16 in. in diameter.

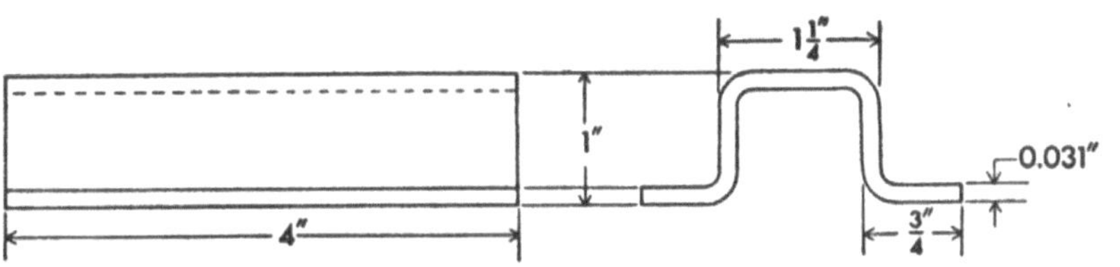

Fig. 275. A stainless-steel, hat-section channel.

Problem 4. Using stainless steel having a thickness of 0.046 in., prepare two flanges having the dimensions shown in Figure 274 and fasten them together, using 1/16 in. rivets.

Problem 5. From 18–8 sheet having a thickness of approximately 0.031 in., form the hat-section channel shown in Figure 275.

XXI ASSEMBLY

An airplane is not built as a whole, but is built in parts called subassemblies which are fastened together in the final assembly to form the whole airplane. Assembly forms a considerable part of the duties of the aircraft, sheet-metal worker. After the individual parts are formed, they are fastened together in subassemblies which may consist of a rudder, an aileron, a wing, or a fuselage part. These small subassemblies are then assembled as a group, such as the tail group, the center-section group, landing-gear group, or, on large airplanes, parts of the fuselage, such as the nose section, cabin section, or complete wings. These main subassemblies are then assembled to form the complete airplane on the final assembly line.

Assembly consists of fastening the various parts and subassemblies into place. Several different kinds of fasteners are used. Rivets are one of the most important of the fasteners used in assembly. Riveting is always an assembly process. Some parts may be fastened by means of bolts, screws, self-tapping screws, machine screws, or studs. Many different kinds of nuts are used, such as wing nuts, check nuts, self-locking nuts, nut plates, and elastic stop nuts. When bolts are used, they usually have to be safetied by means of cotter keys or safety wire and may be fitted with washers to keep the nut from injuring the structural material. Cowlings are usually fastened into place by special cowling fasteners so that the cowling may be removed or opened for service and inspection. Inspection plates may be fastened with screws which screw into nut plates. Temporary fasteners are often used to hold the parts in place during assembly. These fasteners may consist of clamps, spring clamps, parts of a jig, or Cleco clamps. Many small subassemblies are fastened by spot welding, as shown in Figures 277 and 278.

In addition to the various hand tools, assembly requires a number of special jigs. Jigs must be constructed with extreme accuracy. The tolerance of most jig structures is not more than $\frac{1}{100}$ in. Jigs vary in size

Fig. 276. Assembling an airplane door by spot welding. (Courtesy Aluminum Company of America)

from those used to assemble an aileron, rudder, or fuel tank to those which determine the shape of the entire airplane. These jigs are known as wing jigs, fuselage jigs, etc., being named after the part formed in them. Some jigs resemble check blocks, over which various parts are assembled.

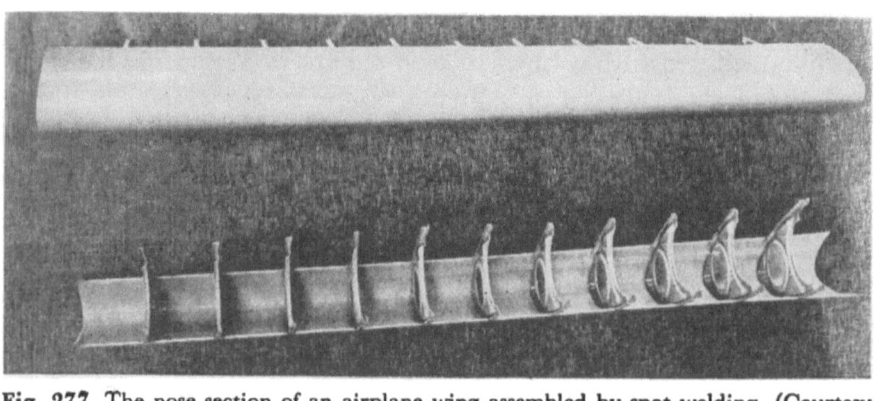

Fig. 277. The nose section of an airplane wing assembled by spot welding. (Courtesy Aluminum Company of America)

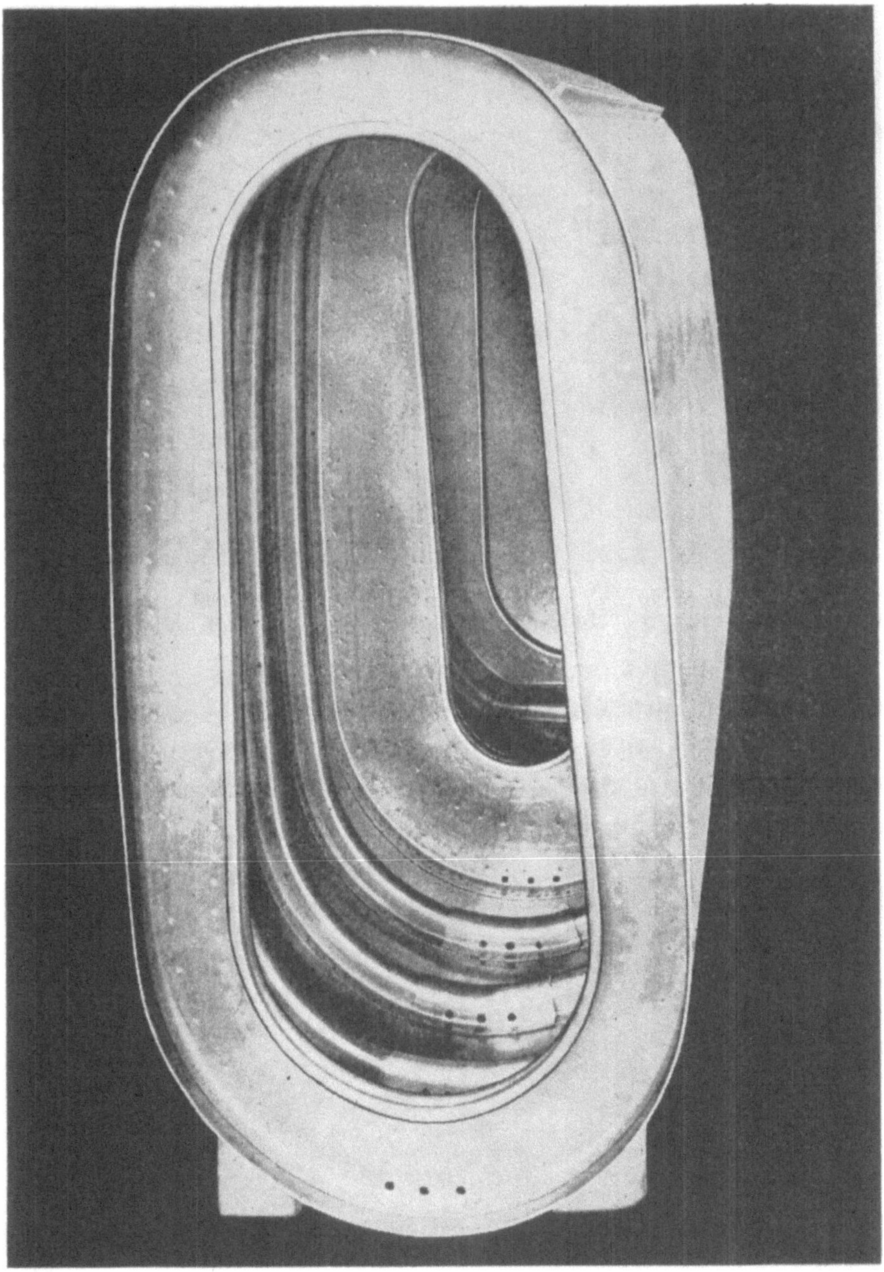

Fig. 278. A fuel tank constructed by spot welding. (Courtesy Aluminum Company of America)

Fig. 279. A wing mounted on a jig for assembly. Note the diagonal rib construction. (Courtesy Engineering Research Corporation)

Jigs should be so formed that the parts fit readily into place and are easily removed when assembly is complete. Many jigs are so formed that the assembly may be rotated to allow easy access to all parts. Figure 279 shows an airplane wing mounted on such a jig. This wing shows diagonal rib construction.

Fig. 280. Mass production parts ready for subassembly on an automatic riveting machine. (Courtesy Douglas Aircraft Company, Inc.)

ASSEMBLY

All jigs must be designed for the particular part being assembled. Fuselage jigs are usually heavy, rigid structures designed to hold bulkheads, stringers, cross members, and other structural parts in place while being assembled. Many fuselage jigs are so designed that the skin is fastened to the structural framework while still in the jig.

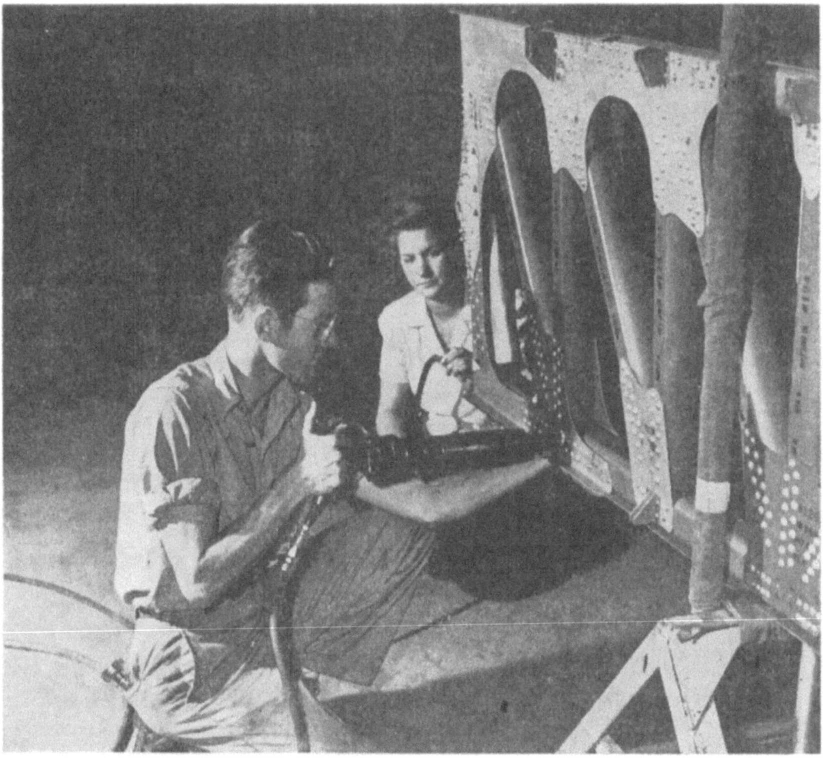

Fig. 281. Riveting a subassembly. (Courtesy Douglas Aircraft Company, Inc.)

Light airplanes usually have the wing assembled in one panel. The jig is constructed of two uprights connected with two cross members. The cross members locate the front and rear spars. The fuselage fittings in the root of the wing are usually used for bolting points. The ribs slide over the spars and are fastened into place. Wings are usually assembled with the spars horizontal and the wing surfaces vertical. This allows work on both the top and bottom surfaces of the wing at one time. Large wings are built in sections, one of which may consist of a center section. To the center section may be fastened the engine nacelles, engine mounts, and landing-gear parts. Center sections often contain the fuel tanks. The outer portion of the wing may be assembled in two parts consisting of the main panel and the tip. Some large airplanes have a root panel which is fastened directly to the fuselage. Figure 284 shows a fuselage being lowered onto the center section during the assembly of a large, four-motored airplane.

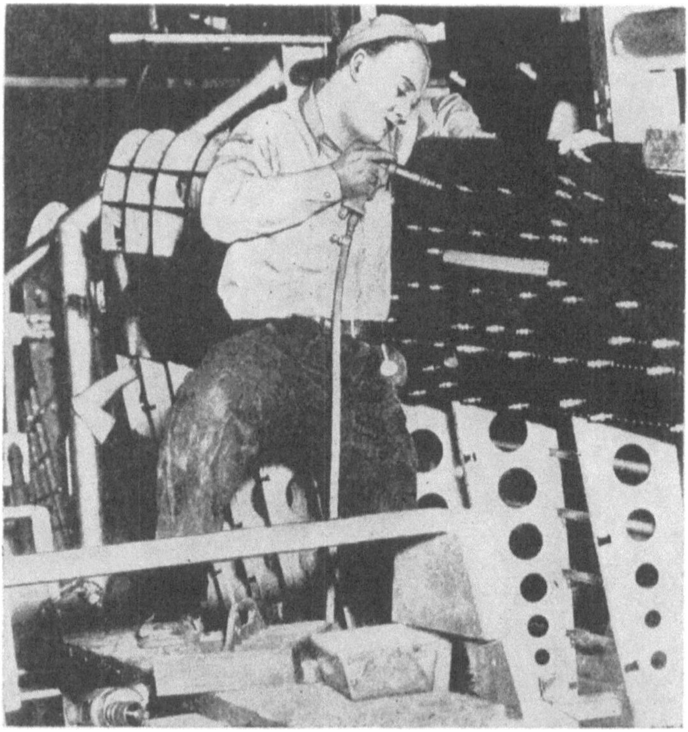

Fig. 282. Riveting a wing covering in place. Note the jig and the large number of Cleco clamps holding the skin in place. (Courtesy The Cleveland Pneumatic Tool Company)

After the framework of subassemblies has been completed, the skin is riveted into place. The sheet-metal worker must be familiar with the working characteristics of all the kinds of sheet metal used for skins. In fitting the part, a considerable amount of handwork which cannot be shown on the drawings is often necessary. The skin fitter must use good judgment and have a high degree of mechanical skill. Considerable experience is necessary to perform this operation properly.

The metal skin of an airplane is usually of thin material which must be carefully handled to prevent damage. The covering should be applied with the grain of the metal extending in the direction of greatest stress. If the grain is difficult to see, it may be brought into view by wiping the sheet with nitrate dope thinner. On wings and control surfaces, the grain should run parallel to the spars. The grain, when possible, should be arranged to run lengthwise on the fuselage. It is the responsibility of the skin fitter to check the grain on all parts of the skin which he applies. Many times a part needs additional forming to prevent

Fig. 283. Wing assembly showing jigs which allow the wing to rotate. (Courtesy The Cleveland Pneumatic Tool Company)

the skin from wrinkling or buckling. It may be necessary to remove a part which has been fitted into place and send it back to be reformed. However, the skin fitter should be able to make minor adjustments. If the part is undersized, it must be scrapped. When the fit is satisfactory, a helper, working from the inside of the sheet, drills the necessary rivet holes. Pilot holes are usually drilled in the framework before the skin is fitted. These pilot holes in the structural parts to which the skin is to be fastened are usually drilled with a No. 50 drill. No. 30 is a common size of drill for skin rivets. Pilot holes may be drilled from the inside of the skin and then reamed from the outside with the proper size of drill before riveting. The portion of skin being drilled is held in place by padded wooden blocks. The first holes are usually drilled near the center of the sheet being applied. As soon as these holes are drilled, skin fasteners, such as Cleco clamps or machine screws having fiber washers inserted under the heads to protect the skin, are put in place to secure the sheet. Sheet-metal screws should not be used to hold the sheet in place after the holes have been drilled to full size. The holes are drilled, and the fasteners inserted, working outward from the center of the sheet. Any wrinkles that form must be worked out. Wrinkles and bulges may form oil cans. The skin fitter can test for this condition by pressing

Fig. 284. A fuselage being lowered into place on a one-piece inner-wing or center section. (Courtesy Douglas Aircraft Company, Inc.)

on the sheet and determining whether it will spring back and forth in the same manner as the bottom of an oil can. Oil cans must always be removed, as vibration may cause failure of the parts.

After the holes are drilled and the skin is properly fitted, all burrs should be removed from the skin and structural members. If the skin is to be flush riveted, any one of a number of different methods may be used. The structural part to which the sheet is riveted may have the holes countersunk or dimpled. Dimples may be formed by a machine or by hand tools. On thin sheet the dimple may be formed by the rivet during the driving operation. The skin is then put in place and tried

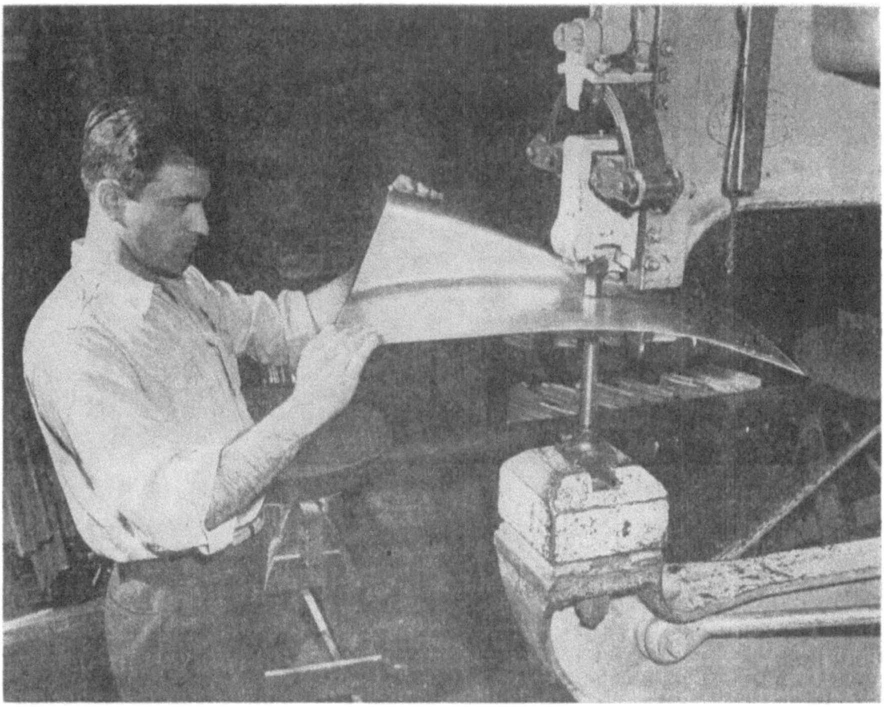

Fig. 285. Final finishing is often accomplished with a Yoder hammer. (Courtesy The Glenn L. Martin Company)

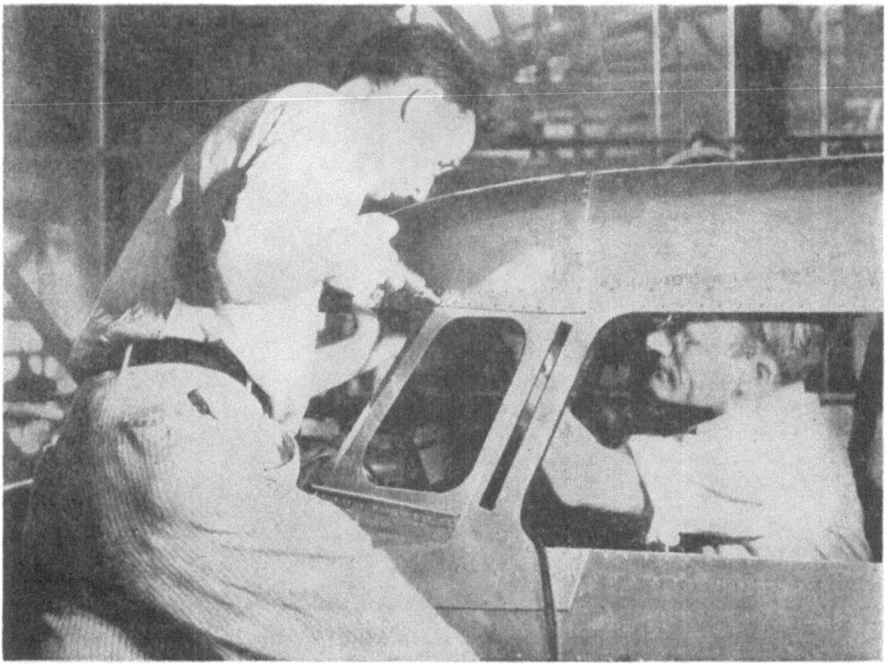

Fig. 286. Assembling a cabin section. (Courtesy Aluminum Company of America)

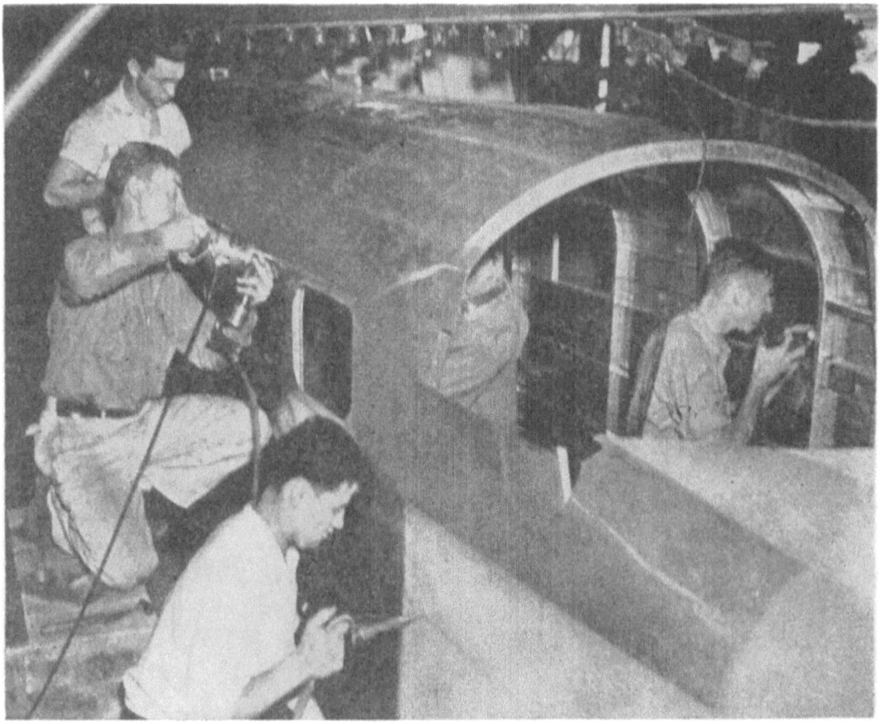

Fig. 287. Fuselage assembly. The worker at the right is bucking a rivet; the worker at the left center is drilling rivet holes; and the worker in the foreground is setting rivets with a rivet gun. (Courtesy Aluminum Company of America)

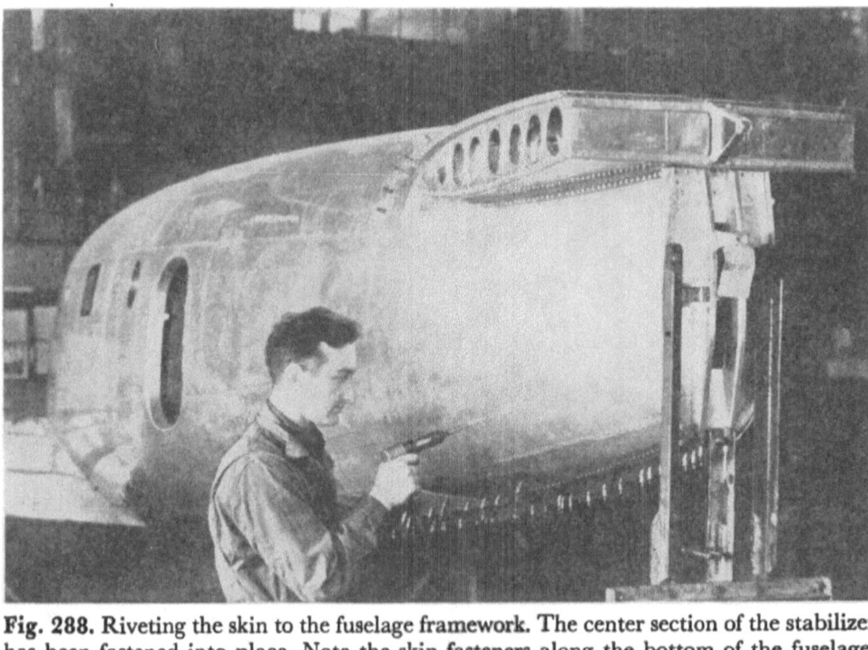

Fig. 288. Riveting the skin to the fuselage framework. The center section of the stabilizer has been fastened into place. Note the skin fasteners along the bottom of the fuselage. (Courtesy The Cleveland Pneumatic Tool Company)

Fig. 289. The assembly line in a large aircraft factory. (Courtesy Douglas Aircraft Company, Inc.)

Fig. 290. Final assembly. (Courtesy Douglas Aircraft Company, Inc.)

for proper fit. When the skin fits, the proper number of temporary fasteners to hold the sheet firmly in place is inserted through the rivet holes. Riveting is started near the center of the sheet, working outward

Fig. 291. Part of the skin has been left off the underside of a wing to allow installation of tubing and accessories. (Courtesy Aluminum Company of America)

in the same manner in which the holes were drilled. Any tendency to buckle or deform should be immediately reported to the supervisor. It is often necessary that various parts of the skin be left off temporarily to allow for installation of pipes, conduits, fittings, and accessories.

XXII SHEET–METAL REPAIRS

All repairs to certificated aircraft must either be made in a manner approved by the Civil Aeronautics Administration or by an inspector of the Administration.

The Administrator of the Civil Aeronautics Administration has designated who may perform approved repairs: (1) the manufacturer of the aircraft; (2) an approved repair station; and (3) a certificated aircraft mechanic.

Aircraft repairs are divided into two classes. Minor repairs consist of repairs to non-structural members. Major repairs are repairs made to parts of vital importance to the airworthiness of an aircraft and include the replacement, strengthening, reinforcement, and splicing of structural members, such as fuselage longerons, landing-gear parts, and fittings.

In making any repairs, approved workmanship and technique must be used. All materials used must be of equal or better quality than the original material, and the repair must always equal or exceed the original strength.

Emergency repairs may be of a temporary nature and must be replaced by permanent repairs as soon as possible. All permanent repairs are made in such a manner that they are expected to become a permanent part of the aircraft and should, therefore, be made at least as strong as the original part and finished in an approved manner. All permanent repairs must be made in accordance with the Civil Air Regulations as established by the Civil Aeronautics Administration.

Extensive repairs to sheet-metal structures should be made by the manufacturer or an approved repair station rated for this type of work. Only a certificated mechanic thoroughly experienced in the work should attempt repairs to metal aircraft structures.

Unconventional fastenings, such as self-tapping screws, driving screws, "lock-scrus," or any other fastening, unless originally used by the manufacturer, may be used only with the approval of the Civil Aeronautics Administration. No aluminum alloy bolts having a diameter of less than ¼ in. should be used in the main structure of an airplane. Aluminum alloy bolts may be used only where it is not necessary to remove them in making inspections, or for the purpose of maintenance and service. Aluminum alloy nuts may not be used on seaplanes. They may be used in land planes on steel bolts in shear, providing the bolts are cadmium plated. Self-locking nuts may only be used in main structures when approved by the Civil Aeronautics Administration. No oversized hole should be drilled where there is any danger of decreasing the original strength of the structure. All repairs to such parts as wing-spar cap strips, fin, fuselage, wing longitudinal stringers, or other highly stressed members should be made in accordance with factory recommendations and the approval of the Civil Aeronautics Administration.

When disassembling in preparation for repairs, all parts of the structure should be supported to prevent damage. Rivets should be removed by drilling through the head and driving out with a punch, while the material is supported to prevent damage. All rivets near a damaged area should be carefully inspected for partial failure. Sample rivets should be removed from nearby areas to determine whether or not the holes have become elongated or the rivets have become partially sheared.

Do not attempt to heat-treat parts after riveting as this will cause warping. Riveted assemblies should not be heated in a salt bath. Small amounts of salt will remain in cracks and crevices and lead to corrosion. A17S–T rivets may be driven as received but 17S–T and 24S–T should be heated before riveting. Some of the small-sized rivets may be driven as received, but it is better practice always to reheat before driving. Rivets should be replaced with a size equal to the original rivet. If it is necessary to enlarge the hole to accommodate the next larger size of rivet, it may be necessary to obtain Civil Aeronautics Administration approval. Rivets should be properly spaced and never be used in a structure where they will be subjected to direct tension. Direct tension tends to pull the heads off. 17S–T rivets of 3/16-in. diameter or less and 24S–T rivets of 5/32-in. diameter or less may be replaced with A17S–T rivets for general repairs. A17S–T rivets must be 1/32 in. greater in diameter than the rivets they replace. Hollow rivets or rivets of decidedly different types may not be substituted for solid rivets without the approval of the Civil Aeronautics Administration. Care should be taken in this type of repair that the rivets in the adjacent structure are not loosened due to the vibrations set up by riveting. All nearby rivets should be carefully examined after the repair has been completed. Where possible, a rivet squeeze should be used when making repairs. In riveting through tubes or hollow structures, the rivet should be driven only enough to form a small head. To attempt to form a standard head would cause the rivet to buckle within the tube.

All material used in making repairs should, whenever possible, be the same as the original material. Any material substituted must be of the same type of alloy. 24S–T may be substituted for 17S–T, but 17S–T should not be substituted for 24S–T unless a heavier gauge is used. All repairs to 24S–RT alloy structures, whenever possible, should be made with the same material as the original.

Care must be taken to determine that the material being used has been properly heat-treated. When doubt exists as to the composition of a material, proper tests should be made.

Parts which have been formed in the annealed condition may have bends which are impossible to make in material in the heat-treated condition. Material should not be heated to assist in forming unless reheat treatment restoring its original properties can be carried out.

Bend lines should be made to lie as nearly as possible at 90° to the grain of the metal. A large bend radii should be allowed for material in the T condition. All sheet-metal parts, which are not to be painted or otherwise finished, should be made from aluminum-coated (alclad) material. All parts used in repairs should be carefully checked for corrosion, pits, kinks, tool marks, scratches, cracks, or any other defects before being installed.

Damaged aluminum alloy ribs of either the built-up or stamped-sheet type may be repaired by suitable reinforcements. This type of repair is acceptable in small and medium-sized airplanes. Unusual types of repairs require the approval of the Civil Aeronautics Administration.

Small holes or cracks in the metal skin, which do not involve damage to structural members, may be patched by covering the hole with a metal plate. These patches must be of such size that they allow for proper edge distance and rivet spacing. In case the damage to the metal skin is extensive or the damaged area overlaps structural members, such as bulkheads and stiffeners, the entire damaged sheet should be removed. In replacing a new sheet, the seams, rivet pattern, and rivet size should be the same as the original. If the original seams are not alike, the stronger of the seams should be copied.

Structural members which have been slightly bent may be straightened cold. However, after straightening, the area should be examined with a magnifying glass for injuries to the material. The straightened parts should then be reinforced. The reinforcement may be less than the original strength, depending upon the amount of damage, because the part retains some of its original strength. If any strain cracks or decided indications of failure are present, the reinforcement should be made equal in strength to the original part. The reinforcement should be made by following the manufacturer's "Administrator of Civil Aeronautics Approved" recommendations. In general, the attachment of all reinforcements should extend beyond the damaged area and be fastened to sound metal.

Local heating to assist in bending or any other forming should not be used on aluminum heat-treated alloys. It is sometimes permissible to use a torch with a large, soft flame to anneal slightly nonheat-treated alloys of the work-hardened type. This should never be done if it is possible to anneal in a furnace or a bath. The material should not be heated to a temperature above the minimum charring temperature for a resinous pine stick. Splices to stringers and flanges should be made in accordance with the manufacturer's recommendations. These recommendations are usually contained in an "Administrator of Civil Aeronautics Approved" repair manual. Stringers are designed to carry both tension and compression forces. When splicing members of this type, the net cross-section area of the spliced member should, in general, be greater than the area of the section of the part spliced. Standard spacing and edge distance of rivets in splices should be followed. When repairing cracks, small holes of 3/32 in. or 1/8 in. should be drilled at the extreme end of the crack to prevent further cracking. All reinforcements should be designed to carry the stresses across the damaged portion and stiffen the joints. When repairing welded or riveted tanks, sealing compounds which are insoluble in gasoline or oil should be used to make a tight joint. All traces of flux should be carefully removed after welding.

Aluminum alloy tubing should not be annealed after forming or at overhaul periods, as is required for copper tubing. Spot-welded structures or assemblies should be repaired at an approved station or by the factory unless approved repairs may be accomplished by substituting rivets for spot welds. Castings, as a rule, should be replaced, and no attempt should be made to repair such parts. Unless aluminum-coated materials are used, protective coatings should be applied. Where possible, anodizing and proper priming should be carried out on newly repaired parts. Aluminum alloys in contact with wood or dissimilar materials should be protected from corrosion by one or more coats of zinc chromate primer, also the wooden parts should be made thoroughly waterproof.

The general methods of making repairs to sheet-metal airplane structures may be found in Civil Aeronautics Manual 18. Any repairs not made in accordance with this manual must be approved by the Civil Aeronautics Administration.

INDEX

The Aviation Collection by Sportsman's Vintage Press

www.SportsmansVintagePress.com

Aircraft Construction Handbook	by Thomas A. Dickinson
Aircraft Sheet Metal Work	by C. A. LeMaster
The Aircraft Apprentice	by Leslie MacGregor
Aircraft Woodwork	by Col. R. H. Drake
Aircraft Welding	by Col. R. H. Drake
Aircraft Sheet Metal	by Col. R. H. Drake
Aircraft Engines	by Col. R. H. Drake
Aircraft Electrical and Hydraulic Systems, and Aircraft Instruments	by Col. R. H. Drake
Aircraft Engine Maintenance and Service	by Col. R. H. Drake
Aircraft Maintenance and Service	by Col. R. H. Drake

Made in United States
Troutdale, OR
07/27/2026

51644927R00139